Studies in Applied Philosophy, Epistemology and Rational Ethics

Volume 20

About this Series

Studies in Applied Philosophy, Epistemology and Rational Ethics (SAPERE) publishes new developments and advances in all the fields of philosophy, epistemology, and ethics, bringing them together with a cluster of scientific disciplines and technological outcomes: from computer science to life sciences, from economics, law, and education to engineering, logic, and mathematics, from medicine to physics, human sciences, and politics. It aims at covering all the challenging philosophical and ethical themes of contemporary society, making them appropriately applicable to contemporary theoretical, methodological, and practical problems, impasses, controversies, and conflicts. The series includes monographs, lecture notes, selected contributions from specialized conferences and workshops as well as selected Ph.D. theses.

More information about this series at http://www.springer.com/series/10087

Lorenzo Magnani · Ping Li
Woosuk Park
Editors

Philosophy and Cognitive Science II

Western & Eastern Studies

Springer

Editors
Lorenzo Magnani
Department of Philosophy and
Computational Philosophy Laboratory
University of Pavia
Pavia
Italy

Woosuk Park
School of Humanities and Social Sciences
Korea Advanced Institute of Science and
Technology
Daejeon
Korea, Republic of (South Korea)

Ping Li
Department of Philosophy
Sun Yat-sen University
Guangzhou
China

ISSN 2192-6255 ISSN 2192-6263 (electronic)
Studies in Applied Philosophy, Epistemology and Rational Ethics
ISBN 978-3-319-18478-4 ISBN 978-3-319-18479-1 (eBook)
DOI 10.1007/978-3-319-18479-1

Library of Congress Control Number: 2015939807

Springer Cham Heidelberg New York Dordrecht London

Printed on acid-free paper

Springer International Publishing AG Switzerland is part of Springer Science+Business Media
(www.springer.com)

Preface

Today, the relationships between Asia and the Western world make headlines only when they concern economic deals, folk-ideological confrontations, or divergent ideas on how to solve international crises. The cultural and, more specifically, academical links are frequently disregarded. This book aims at being an argument against such systematic lack of interest for the results of collaborations between Western and Eastern intellectuals and academics: what emerges from the juxtaposition of papers of different geo-cultural origins—but dealing with the same issues—is sometimes a novel approach, which takes advantage of the multifaceted sensibilities inherited by the scholarly legacies that contributed to the debate. This volume is a collection of selected papers that were presented at the international conference *Philosophy and Cognitive Science* (PCS2013), held at Sun Yat-sen University, Guangzhou, P.R. China, in November 2013 (chairs Lorenzo Magnani and Ping Li) and at the International Workshop *Visual Abduction or Abductive Vision*? held at KAIST (Korea Advanced Institute of Science and Technology), Daejeon, South Korea, in October/November 2013 (chair Woosuk Park).

The papers by Athanassios Raftopoulos "Reframing the Problem of Cognitive Penetrability," Xiang Chen "The Emergence and Development of Causal Representations," Luigi Pastore, Sara Dellantonio, Claudio Mulatti, and Remo Job "On the Nature and Composition of Abstract (Theoretical) Concepts: The X-Ception Theory and Methods for Its Assessment," Selene Arfini and Lorenzo Magnani "An Eco-Cognitive Model of Ignorance Immunization," Woosuk Park "Towards a Caricature Model of Science," and Lorenzo Magnani "Violence and Abductive Cognition Epistemology and Ethics Entangled" were presented at PCS2013. The papers by Lorenzo Magnani "Understanding Visual Abduction. The Need of the Eco-Cognitive Model," Cameron Shelley "Biomorphism and Models in Design," Jeongmin Lee, "The Correspondence Principle, Formal Analogy, and Scientific Rationality," Jun-Young Oh, YooShin Kim, Chun-Hwey Kim, Byeong-Mee Min, Yeon-A Son "Understanding Galileo's Inquiries about the Law of Inertia," Athanassios Raftopoulos "Abductive Inference in Late Vision," and Woosuk Park "From Visual Abduction to Abductive Vision" were presented at the KAIST Workshop on abduction.

Previous volumes prepared the basis for the realization of PCS2013 and of KAIST Vision Workshop, as meetings explicitly devoted to the conjunction of Western and Eastern studies. These volumes also originated from international joint research projects, which succeeded in establishing a first relationship between the two worlds in the area of philosophy and cognitive science. *Model-Based Reasoning in Scientific Discovery*, edited by L. Magnani, N.J. Nersessian, and P. Thagard (Kluwer Academic/Plenum Publishers, New York, 1999), based on the papers presented at the first "model-based reasoning" international conference, held at the University of Pavia, Pavia, Italy in December 1998, has been translated into Chinese, China Science and Technology Press, Beijing, 2000. *Abduction, Reason, and Science* by L. Magnani was translated by Dachao Li and Yuan Ren and published by Guangdong People's Publishing House, Guangzhou, in 2006. Other volumes, *Science, Cognition, and Consciousness*, edited by P. Li et al. (JiangXi People's Press, Nanchang, China, 2004, published in Chinese and English), *Philosophical Investigations from a Perspective of Cognition*, edited by L. Magnani and P. Li (Guangdong People's Publishing House, Guangzhou, China, 2006, published in Chinese), *Model-Based Reasoning in Science, Technology, and Medicine*, edited by L. Magnani and P. Li (Springer, Berlin/New York, 2007), derived from the following previous conferences: "Model-Based Reasoning in Science and Medicine" (MBR06_CHINA, held at Sun Yat-sen University, Guangzhou, China, July 2006), the first "Philosophy and Cognitive Science" international conference (PCS2004, held at Sun Yat-sen University, Guangzhou, China, June 2004) and the second "Philosophy and Cognitive Science" international conference (PCS2011, held at Sun Yat-sen University, Guangzhou, China, May 2011).

The presentations given at the Guangzhou and Daejeon meetings addressed various recent topics at the crossroad of philosophy and cognitive science, especially taking advantage of both Western and Eastern research. The selected papers contained in the proceedings mainly focus on the following areas: abductive cognition, visualization in science, the cognitive structure of scientific theories, the nature and functions of models, scientific representation, mathematical representation in science, model-based reasoning, analogical reasoning, moral cognition, cognitive niches, and evolution. The various contributions of the book are written by interdisciplinary researchers who are active in the area of philosophy and/or cognitive science.

The editors wish to express their appreciation to the members of the Scientific Committee of PCS2013 for their suggestions and assistance: Xiang Chen, Department of Philosophy, California Lutheran University, Thousand Oaks, CA, USA—Roman Frigg, Department of Philosophy, Logic and Scientific Method, London School of Economics and Political Science, London, UK—Albrecht Heeffer, Center for Logic and Philosophy of Science, Ghent University, Ghent, Belgium—Remo Job, Department of Psychology and Cognitive Science, University of Trento, Rovereto (TN), Italy—Ping Li, Department of Philosophy, Sun Yat-sen University, Guangzhou, China—Lorenzo Magnani, Department of Humanities, Philosophy Section, University of Pavia, Pavia, Italy—Woosuk Park,

Korea Advanced Institute of Science and Technology, Guseong-dong, Yuseong-gu Daejeon, Republic of Korea—Mauricio Suárez, Department of Logic and Philosophy of Science, Faculty of Philosophy, Complutense University of Madrid, Madrid, Spain—Ryan D. Tweney, Department of Psychology, Bowling Green State University, Bowling Green, OH, USA—Xiaolong Wan, Department of Philosophy, Huazhong University of Science and Technology, Wuhan, Hubei, China—Guolin Wu, Center for Philosophy of Science and Technology, South China University of Technology, Guangzhou, China—Changle Zhou, Department of Cognitive Science, Xiamen University, Xiamen, China—Jing Zhu, Department of Philosophy, Sun Yat-sen University, Guangzhou, China.

Special thanks also go to Tommaso Bertolotti and Selene Arfini for their contribution in the preparation of this volume. The meeting PCS2013 and KAIST Vision Workshop, and thus indirectly this book, was made possible through the generous financial support, respectively, of Sun Yat-sen University, ZhanJiang Chemical Industrial Incorporated Corporation, the MIUR (Italian Ministry of the University), KAIST, South Korea, and University of Pavia, Italy. Their support is gratefully acknowledged. The preparation of the volume would not have been possible without the contribution of resources and facilities of the Computational Philosophy Laboratory and of the Department of Humanities, Philosophy Section, University of Pavia.

Pavia, Italy
Guangzhou, China
Daejeon, Korea, Republic of (South Korea)
January 2015

Lorenzo Magnani
Ping Li
Woosuk Park

Contents

Part I International Conference Philosophy and Cognitive Science (PCS2013)

Reframing the Problem of Cognitive Penetrability 3
Athanassios Raftopoulos

The Emergence and Development of Causal Representations 21
Xiang Chen

On the Nature and Composition of Abstract (Theoretical) Concepts: The X-Ception Theory and Methods for Its Assessment 35
Luigi Pastore, Sara Dellantonio, Claudio Mulatti and Remo Job

An Eco-Cognitive Model of Ignorance Immunization 59
Selene Arfini and Lorenzo Magnani

Towards a Caricature Model of Science 77
Woosuk Park

Violence and Abductive Cognition: Epistemology and Ethics Entangled 95
Lorenzo Magnani

Part II International Workshop Visual Abduction or Abductive Vision? KAIST (Korea Advanced Institute of Science and Technology)

Understanding Visual Abduction: The Need of the Eco-Cognitive Model 117
Lorenzo Magnani

From Visual Abduction to Abductive Vision 141
Woosuk Park

Abductive Inference in Late Vision. 155
Athanassios Raftopoulos

The Correspondence Principle, Formal Analogy, and Scientific Rationality 177
Jeongmin Lee

Understanding Galileo's Inquiries About the Law of Inertia 193
Jun-Young Oh, YooShin Kim, Chun-Hwey Kim, Byeong-Mee Min and Yeon-A Son

Biomorphism and Models in Design 209
Cameron Shelley

Part I
International Conference Philosophy and Cognitive Science (PCS2013)

Reframing the Problem of Cognitive Penetrability

Athanassios Raftopoulos

Abstract I propose a reframing of the problem of cognitive penetrability (CP) that adds to the discussion on whether some cognitive effects on perceptual processing constitute cases of CP a dimension that was initially the main motive for introducing the notion of CP and was later almost abandoned, namely, whether the cognitive effects undermine the epistemological role of perception in grounding perceptual beliefs. I distinguish between intrinsic and extrinsic cognitive effects on perception and argue that intrinsic cognitive effects on perception entail CP, while extrinsic effects entail CP only if they undermine the evidential role of perception in grounding perceptual beliefs. I also explain why the effects of two sorts of "body of knowledge" that are embedded in the visual circuits and guide perceptual processing from within are not cases of CP.

1 Introduction

Under the influence of the work of Hanson (1958) and Kuhn (1962), many philosophers espoused the view that what one thinks determines what they perceive. Perception became theory-laden, conceptually modulated or conceptually non-encapsulated, and cognitively penetrated (CP). (The relations among theory-ladenness, conceptual modulation, and cognitive penetrability are not as clear as they seem. As I have explained (Raftopoulos 2009), however, under some, independent assumptions, the equivalence holds true.)

Seeking to undermine the view that perception is CP, Fodor (1983) argued that perception consists in a series of interconnected modules that are cognitively impenetrable (CI); Fodor posited informational encapsulation as the main characteristic of the perceptual modules. However, not all cases of informational non-encapsulation are cases of CP. The kind of informational exchange that would signify the CP of perception is the flow of conceptual information from cognitive states to perception,

A. Raftopoulos (✉)
Department of Psychology, University of Cyprus, P.O. Box 20537, 1678 Nicosia, Cyprus
e-mail: raftop@ucy.ac.cy

L. Magnani et al. (eds.), *Philosophy and Cognitive Science II*,
Studies in Applied Philosophy, Epistemology and Rational Ethics 20,
DOI 10.1007/978-3-319-18479-1_1

and the use of this conceptual information by the perceptual processes. If some low-level, non-cognitive states affect visual processing by transmitting to it information, visual perception is not informationally encapsulated but is CI. For this reason, I prefer talking of conceptual modulation as the main trait of CP.

The notion of CP was not thoroughly analyzed. With the reinvigorated interest in the nature of perception, the notion of CP became the focus of analysis. Several definitions have been proposed in the literature (Macpherson 2012; Pylyshyn 1999; Siegel 2012; Stokes 2012). All definitions share a common thread; they exclude from instances of CP cases in which the percept is determined through the focus of spatial attention (I discuss only cases of cognitively-driven, endogenous attention), or, in general, cases in which concepts determine indirectly the percept. It is not adequately explained, however, why the indirect effects do not entail CP.

Moreover, philosophers usually discuss the CP of perception as if perception were a unified stage. Perception, however, is not a homogeneous, undifferentiated process. (Among philosophers, Raftopoulos (2009) and Siegel (2012) are sensitive to this distinction.) It consists of two main stages, namely early vision and late vision that are differently affected by cognition through attention. Therefore, discussions on CP/CI should specify the scope of the claim that perception is CP or CI. Moreover, a definition of CP/CI should be able to account for differences (if any) in the CP/CI character of each stage owing to the differences between the ways cognition affects the two visual stages.

In this paper, continuing earlier work (Raftopoulos 2001a, b, 2006, 2009, 2013), I propose a reframing of the problem of cognitive penetrability (CP) that adds to the discussion on whether some cognitive effects on perceptual processing constitute cases of CP a dimension that was initially the main motive for introducing the notion of CP, namely, whether the cognitive effects undermine the epistemological, evidential role of perception in grounding perceptual beliefs.

In the first section, I distinguish between intrinsic and extrinsic cognitive effects on perception (a cognitive effect on perception is extrinsic if it can be mitigated), and claim that intrinsic cognitive effects on perception entail directly the CP of perception. Then, I argue that extrinsic cognitive effects should entail the CP of perception only to the extent that they undermine the evidential role of perception in grounding perceptual beliefs. The motive behind this claim is that the original considerations that gave rise to the discussion concerning the CP of perception were motivated by the view that perception is conceptually modulated and theory-laden. This vitiated the evidential role of perception in evidencing perceptual beliefs. Therefore, if some extrinsic cognitive effects undermine the evidential role of perception, they should be treated as cases of CP, their extrinsic character notwithstanding. In view of these, since late vision is intrinsically affected by cognition, late vision is necessarily CP. Thus, early vision should be the focus of the contemporary discussion about the CP of perception.

In the second section, I examine the sorts of cognitive effects on early vision to determine whether they are intrinsic or extrinsic. I have argued (Raftopoulos 2009) that the available empirical evidence unequivocally suggests that there are no intrinsic cognitive effects on early vision. There are, however, two special sorts of effects on early vision that may be taken as evidence that early vision is intrinsically affected by cognition and is, thus, CP. These are two sorts of 'bodies of knowledge' that are embedded in the visual circuits and guide perceptual processing very early and, consequently, they function within the time scale of early vision. Moreover, they seem to affect early vision intrinsically. The first is the set of general principles that perception employs to solve various underdetermination problems. The second is information that results from perceptual learning and is encoded in the early visual circuits. I examine these two cases and conclude that they are not in effect instances of CP because although some mechanisms causally and intrinsically affect early vision, there is nothing cognitive in these effects and no concepts are involved in the content of early vision states.

There are, however, extrinsic cognitive effects on early vision, such as the effects of spatial attention and the effects of pre-cueing. Following the suggestion that in such cases to determine whether the extrinsic effects constitute cases of CP one should examine whether these effects undermine the evidential role of early vision in grounding perceptual beliefs, I argue that these effects do not undermine the evidential role of early vision and, thus, they do not entail the CP of early vision. This justifies the standard claim to the effect that spatial attention and pre-cueing, which are typical cases of cognitive extrinsic effects on perception, do not constitute cases of CP.

In this paper, I assume that perception consists of a preattentional stage (early vision), and of late vision that is directly modulated by cognitive states through attention.

2 Intrinsic Versus Extrinsic Cognitive Effects on Perception, the Evidential Role of Perception, and Cognitive Penetrability

Before one endeavors to define CP/CI, they should have determined first a set of adequacy conditions that a good definition of CP should fulfill. There are several ways one could go about in this task but I think it promising to start with factors that underlie most discussions of CP/CI. The first is that talk about the CP of perception is a talk about cognitive influences on perception. The second is that CP is inextricable related to discussions concerning the theory ladenness of perception and the evidential role of perception. I start by discussing the first factor.

There are two ways in which cognitive factors could affect perceptual processing. The first concerns influences that are transmitted in a top-down manner from the cognitive to the visual areas of the brain. For CP to occur there must a top-down flow of conceptual information and, thus, any talk about CP is contingent on the existence of a top-down flow of conceptual information. The empirical literature suggests that most cognitive effects take place through the mediation of cognitively-driven, endogenous attention (Carrasco 2011). The second way concerns cognitive effects on perception that may occur because of the presence of concepts embedded in the visual system itself and not because of any top-down transmission of information from cognitive states. The two different ways in which cognition affects perception impose different demands on the definition of CP/CI, depending on whether cognition affects perception in a top-down manner or from within. I examine in this section the role of top-down effects, and in section two I address the second possibility, as it pertains to early vision.

The top-down effects of attention on perception can be broadly categorized into two classes. The first concerns the effects that emerge as a part of perceptual competition, which, thus, are intrinsic or direct to the perceptual processing. The attentional effects on late vision are a characteristic example of this kind of effects. Emerging from the perceptual competition, these effects influence the perceptual processes by altering the state transformations of which the processes consist. In this case, the cognitive information that allocates attention is also used by the perceptual processes. The second class concerns the effects that do not emerge as part of perceptual competition and, thus, are external to perceptual processing even though they causally affect it. Let us call them 'extrinsic' or 'indirect effects'. Examples of the second kind are the pre-cueing effects and the effects of spatial attention in its capacity as determinant of the locus of focus. The indirect effects do not influence perceptual processing on-line and, thus, the conceptual information carried by the cognitive states is not used by perception. As we shall see, the indirect effects set the initial values of parameters that figure in the equations that express state transformations but other than that they do not affect perceptual processing. Thus, the concepts involved in the affecting cognitive states do not enter the perceptual content. For this reason, extrinsic cognitive effects can be mitigated, a factor that will play a significant role when we will discuss the issue of whether the extrinsic cognitive effects on perception vitiate its evidential role.

2.1 Intrinsic Cognitive Effects and Cognitive Penetrability of Perception

The intrinsic or direct cognitive effects on perception constitute a clear case of CP of perception because they exemplify how perceptual processes are altered as a result of cognitive influences. This sets the first condition that a definition of CP should fulfill. *Visual processes that are intrinsically or directly, in the sense explained above, affected in a top-down manner by cognitive states are CP.* It follows immediately that late vision is CP owing to the fact that late vision is intrinsically modulated by cognition.

2.2 Extrinsic Cognitive Effects and Cognitive Penetrability: The Evidential Role of Perception

In the case of the extrinsic cognitive effects, the perceptual processes do not use the concepts involved in the contents of the cognitive states that indirectly affect perception. This leaves open the issue of whether these effects should be construed as cases of CP. It is arguable that even if the cognitive effects are extrinsic, still, the concepts contained in the cognitive states involved influence perception rendering it CP. I think that to answer this question, one should take recourse to the motives underlying the introduction of the problem of the CP of perception. It is, thus, time to bring into the discussion the second factor that should be considered when determining whether some cognitive effects on perception constitute cases of CP. Hanson (1958), Kuhn (1962), Churchland (1988) and others interpreted findings in psychology and neuropsychology as showing that cognitive states involving propositional (conceptual) contents affect perception. This was used as a springboard to mount an attack on the received view in the philosophy of science that there is a theory neutral observational basis on which a rational choice for empirical adequacy between competing theories could be made. Perception becomes theory–laden and the choice between two alternative and mutually exclusive theories cannot be based on empirical testing because being in different conceptual frameworks means that one cannot see others' data and, most importantly, these data, or rather, the experimental observations, are already interpretations made under the influence of the two alternative theories. From this ensues the incommensurability thesis that bars communication across paradigms; perceptions are modulated by theoretical commitments and proponents of different paradigms perceive different worlds.

Sellars (1956), on his part, sought to undermine one of the tenets of classical empiricism, to wit, the view that perception functions independently of concepts and deliver to us the world in its own guise without any conceptual influences. This 'given' can be used as a neutral basis on which to determine the adequacy of both perceptual beliefs and scientific theories.

The thread that connects these theses is the view that perception cannot play the epistemological role traditionally assigned to it by empiricism because it does not provide a neutral ground on which to decide which of our cognitive schemes is true or false. The most important consequence of the CP of perception concerns the epistemological role of perception as evidence in grounding perceptual beliefs, a hypothesis that is reinforced by the recent resurgence of the discussion about the epistemological repercussions of CP. Therefore, if some extrinsic cognitive effects undermine the evidential role of perception, they should be treated as cases of CP, their extrinsic character notwithstanding.

In view of these considerations, a definition of CP/CI should *warrant that if perception (or a stage of it) is indirectly conceptually influenced in a way that renders it unfit to play the role of a neutral epistemological basis, by vitiating its justificatory role in grounding perceptual beliefs, perception (or a stage of it) is CP. If perception (or a stage of it) is indirectly conceptually influenced in a way that does not preclude it from playing this role, it is CI.* The motive underlying the second part of this condition is that if the extrinsic cognitive effects on perception fail to undermine perception's evidential role, they are epistemologically speaking uninteresting and, certainly, discussions on the CP of perception purport to make some epistemological interesting claims. Let us call this condition on CP, the epistemological criterion for CP.

Late vision, by being intrinsically affected by cognition, is clearly theory-laden and conceptually influenced. Therefore, the role of the contents of late vision states in evidencing perceptual beliefs is vitiated on account of the fact that some among the justifying contents of its states are formed through processes that have been affected by these same beliefs that they purport to support epistemically. It follows that perceptual states that satisfy the first criterion, that is, states that are intrinsically affected by cognitive states, also meet the second condition, that is, their evidential role in grounding perceptual beliefs is vitiated.

The first condition leaves early vision as the only viable candidate for being CI since the other stage of visual processing, namely, late vision is intrinsically affected by cognitive states and, thus, is CP. To determine whether early vision is CI, one has to examine the attentional effects on early vision.

3 Early Vision

3.1 Intrinsic Cognitive Effects on Early Vision

If there were intrinsic cognitive effects on early vision, then, *per* the first condition, early vision would be CP. It has been argued (Pylyshyn 1999, 2003; Raftopoulos 2009) that early vision is not directly affected by cognition. This, at a first glance, contradicts evidence that spatial attention affects intrinsically early vision, as evidenced by the modulation of the P1 and N1 ERP components. Were the intrinsic effects of spatial attention on early vision cognitive, early vision would be CP. Some of the effects of spatial attention are intrinsic in that they emerge from the perceptual competition, in that spatial attention enhances or filters out information from the onset of perceptual processing.

I have argued (Raftopoulos 2009), however, that the modulation of the P1 and N1 ERP components is entirely bottom-up and is not affected by any top-down cognitive influences; the modulation of P1 and N1 is an exogenous, data-driven effect and is independent of concepts. Since for CP to obtain the effects on perception should be cognitive, the modulatory P1 and N1 effects of spatial attention on early vision do not signify that early vision is CP. (In contradistinction, the N2 effect of spatial attention is a cognitive effect (Raftopoulos 2009) and, thus, entails that the stage in which this effect takes place is CP. This stage is late vision.) It follows that there seem to be no intrinsic cognitive effects on early vision due to spatial attention.

Independent of the existence and role of top-down cognitive attentional effects on early vision, however, there is also the possibility that early vision is intrinsically affected by cognition and, thus, is conceptually influenced and theory-laden in the second way we have discussed. That is, early vision may be conceptually influenced not because of some top-down transmission of information but because some concepts are embedded in its neural circuits. There are two sorts of effects on perceptual processing, which, because they affect early vision and determine its output and because they play an active role in the perceptual competition, may be taken to entail that early vision is intrinsically affected by cognition. The first concerns the role of some 'principles' that express some 'bodies of knowledge', which constrain perceptual operations. The second concerns perceptual learning and the way it affects perceptual processes. Both sorts of effects have the characteristic that they are realized by mechanisms embedded in early vision. If it turns out that owing to the existence of concepts embedded in the circuits of early vision these effects are cognitive, early vision is CP from the within. So, let us examine them.

3.1.1 Operational Constraints in Perception

Visual processing does not function free of any internal restrictions, but is constrained and modulated at every level by certain principles, or operational constraints that embody certain principles. Because the retinal image underdetermines both the distal object and the percept, perception would not be feasible if the processing of information was not constrained by 'assumptions' that substantiated reliable generalities about the physical world and its geometry. This point is made by most computational theories (Marr 1982; Fodor 1983), and there is evidence that physiological visual mechanisms implement such constraints in their design, from cells for edge detection to mechanisms implementing the epipolar constraint. The constraints are described by many vision theorists (see, for example, Marr (1982, p. 185)) as 'hard-wired' into the visual system and as reflecting 'some kind of a statistical rule of the universe'. (Note that Fodor (1983) uses the term 'hard-wired' to describe modular processing. However, Fodor also thinks that constraints could be implemented within modules by sentence-like structures, a point with which, as we shall see, I disagree.)

Burge (2010) calls the constraints 'formation principles'. I (Raftopoulos 2001a, 2009) has called them 'operational constraints' on vision, which is the term I adopt here. Operational constraints reflect general or higher-order physical regularities that govern the behavior of objects in our world and the geometry of the space around us. Through causal interaction with the environment over the course of evolution, they have gradually been incorporated into the perceptual system. They allow us to perceptually lock onto medium size lumps of matter in the world by providing the discriminatory capacities necessary for the individuation and tracking of objects in a bottom-up, nonconceptual way (Raftopoulos 2009), and they allow perception to generate perceptual states that present worldly objects as cohesive, bounded solids, and as spatio-temporally continuous entities (Spelke 1988).

The constraints are not available for introspection, function outside the realm of consciousness, and cannot be attributed as acts to the perceiver. One does not believe implicitly or explicitly that an object moves in continuous paths or that it is rigid, though they use this information to parse objects. These constraints are not perceptually salient but one must be 'sensitive' to them if they are to be described as perceiving their world. The constraints constitute the *modus operandi* of the perceptual system and not a set of rules used by the perceptual system either as premises in inferences or as rules in inferences; the *modus operandi* of the visual system consists of operations determined by laws describable in terms of computation principles. They are reflected in the functioning of perception and can be used only by it, whereas "theoretical" constraints are available for a wide range of cognitive tasks. These constraints cannot be overridden since they are not under the perceiver's control; one cannot substitute them with another body of constraints even if they know that they lead to errors.

Haugeland (1998) argues that we share with non-concept possessing creatures various innate "object-constancy" and "object-tracking" mechanisms that automatically 'lock onto' medium sized lumps. These mechanisms provide the discriminatory capacities necessary for the individuation and recognition of objects in a bottom-up, nonconceptual way. Haugeland (1998, pp. 261–261) claims that the objective character of perception, that is the fact that perception is about objects *qua* objects, is due to the role of some normative standards that constitute thinghood. The "constitutive standards for thinghood" are cohesiveness and compatibility. These standards are in fact results of the operational constraints on perception that I discussed above.

Haugeland (1998, pp. 248–249) claims that neither the perceiver has a discursive cognizance of the standards in some explicit formulation, nor are these standards articulated as rules. Indeed, one could argue that being hardwired, the constraints are not even contentful states of the perceptual system, or, if they are, the contents are not conceptual, propositionally structured contents that could constitute some theory or other. Let me explain this. A neural state is formed through the spreading of activation and its modification as it passes through the synapses. The hard-wired constraints specify the processing, i.e., the transformation from one state to another, but they are not the result of this processing. They are computational principles that describe transitions between states in the perceptual system, and they constitute a computational processor. Although the states that are produced by means of these mathematical transformations have contents, there is no reason to suppose that the principles that specify the mathematical transformation operations are states of the system and, thus, are represented in the system; this is what the expression *modus operandi* used above purports to convey. That is, even though the perceptual system by using the constraints operates in accord with the principles reflected in them, the perceiver does not represent the constraints.

What, then, about the claim that 'object knowledge' is needed for the filling in that it leads to the construction of the percept? If the operations that effectuate the filling-in are not represented in the system but are performed by hardwired computational processors, is it legitimate to talk about these processors realizing some object knowledge in the form of a set of rules concerning the physical environment and its geometry? This, as a matter of course, depends on what one is willing to count as knowledge.

I stated above that if the operational constraints are not states of the visual system but computational processors, they are not representations or beliefs of any form, either implicit or explicit. (Explicit beliefs are representations that are activated in persons, whereas implicit beliefs are representation that are stored in long-term memory but are not currently activated.) If the constraints are not states of the system, what is the status of the information they contain, that is, of the information included in the regularities about the environment and its geometry that the constraints realize? A first answer is that by not being states, the operational constraints do not have any contents; they are not semantic or mental entities of any kind. To think that they are is a mistake that, as Searle (1995) claims, cognitive scientists often commit. When they have an input and an output state both of which are mental states with representational contents, they tend to think that the processes that connect them are also mental states with some content. There is no reason, however, to assume this as the processes that connect the inputs with the outputs could be non-meaningful, that is, non-contentful, causal connections. According to this view, the function of the operational constraints in perception does not entail that perception is guided by 'object knowledge'.

Other philosophers think that such constraints are states of the system. They use the term 'tacit knowhow' to denote the information carried by states that are built into the system in a way that does not require that the states be represented in any form (Dennett 1983). This tacit knowhow is not represented anywhere in the system and is not a kind of knowledge because if it were we would have to say that birds, in the muscular system of which the laws of aerodynamics are hardwired, know aerodynamics.

There are philosophers who disagree with this view and think that hardwired computational processors realize in a system tacit *knowledge* of a particular set of rules or generalizations (Davis 1995, p. 329). "The rules would not have to be explicitly represented in any representational state of the system. Still less would knowledge of the rules be realized in a state of the same kind as an attitude state." Davis claims that tacit knowledge is not realized by states that are attitude states because tacit knowledge has two main characteristics. First, it is subdoxastic knowledge because tacit knowledge is not inferentially integrated with attitude states and it exists in special-purpose, separate sub-systems (Stich 1978). Second, with attitude states such as beliefs the concepts that are part of the semantic contents of the states must be concepts that are possessed by the believer; the believer should grasp the concepts of which the belief is constituted. This means that the beliefs have their representational contents conceptualized by the believer. The contents of tacit states, however, are not so conceptualized. Moreover, when a person is in a tacit state, they do not have, by being in that state, access to the contents of it. In contradistinction, when a person is in a belief state, they entertain in principle the content of the belief simply by being in that state. It follows that, according to the proponents of tacit knowledge, the operational constraints realize tacit, representational knowledge of the regularities of the physical environment and of its geometry. However, these are not conceptual representations.

Thus, irrespective of the interpretation one favors as to the status of the information realized by the operational constraints, that is, independent of whether one thinks that the constraints are merely causal connectors with no representational contents for the system, or whether this information constitutes some sort of tacit, non-representational knowhow, or whether it is some sort of tacit, representational knowledge, the operational constraints are not rules of inference that are looked-up implicitly or explicitly by the visual system in order to perform its state to state transformations, or premises used in such transformations. Furthermore, the fact that perception relies on such constraints for successful function does not entail that perception is affected from concepts from within As we saw, in any interpretation of the information realized by the operational constraints, this is not conceptual content and, thus, it does not constitute some form of conceptual influence on perception from within. It follows that the existence of the operational constraints that are hardwired in perception does not entail that there is some sort of knowledge that determines perceptual processing. If theories are construed as carriers of knowledge about the world, the operational constraints by themselves do not entail the theory-ladenness of perception.

Thus, Burge (2010) is right to argue that the formation principles do not entail the theory-ladenness of perception:

> For many philosophers, the notion of computational states or explanations is theory-laden in a way that I do not intend. When I call states or explanations 'computational', I do not mean that there are transformations on syntactical items, whose syntactical or formal natures are independent of representational content [of the computed states]. I also do not mean that the principles governing transformation are instantiated in the psychology, or 'looked up', even implicitly in the system … principles governing perceptual transformations … are not the representational content of any states in the system, however unconscious (Burge 2010, p. 95).

3.1.2 Perceptual Learning

One could argue that during our interaction with the world some experiences are learned and form memories that are stored in visual memory and that these memories include certain sensory concepts. In this way, concepts affect perceptual processing rendering it CP from within.

Visual memories affect the way one sees the world. Familiarity with objects or scenes, which is built through repeated exposure to objects or scenes (although some times one presentation is enough), or repetition memory facilitate search, affect figure from ground segmentation, speed up object identification and image classification, etc. (Liu et al. 2009; Peterson and Enns 2005). Familiarity can affect visual processing in different ways. It may facilitate object identification and categorization, which are processes that take time since their final stage occurs between 300–600 ms after stimulus onset, as is evidenced by the P3 responses in the brain, and their earlier stage starts at about 150 ms after stimulus onset (Johnson and Olshausen 2005). Familiarity is in general considered to intervene during the latest stage of object identification and categorization (300–360 ms). These effects are considered to be post-sensory in that they involve the higher cognitive levels of the brain at which semantic information and processing, both being required for object identification and categorization, occur (Delmore et al. 2004).

Familiarity, including repetition memory, may also affect object classification (e.g., whether an image portrays an animal or a face), a process that occurs in short latencies (95–100 ms and 85–95 ms after stimulus onset respectively) (Crouzet et al. 2010; Liu et al. 2009). These effects pose a threat to the CI of early vision since they occur relatively early and cannot be considered post-sensory. The threat would materialize should the classification processes either require semantic information, or require that representations of objects in working memory be activated, since that would entail conceptual involvement. However, researchers agree that the early classification effects result from the feed forward sweep (FFS) and do not involve semantic information, nor do they require the activation of object memories. The main reason for this claim is simple. If they did require any of these two things, they could not be that fast. The brain areas involved are low level visual areas (including the front eye fields) from V1 to V4 (Kirchner and Thorpe 2006), and, a bit more upstream to posterior IT, and lateral occipital complex-LO (Grill-Spector et al. 1998).

The early effects of familiarity may be explained by invoking contextual associations (context spatial relationships) that are stored in early sensory areas to form unconscious perceptual memories (Chaumon et al. 2008), which, when activated from incoming signals that bear the same or similar target-context spatial relationships, modify the FFS of neural activity resulting in the facilitating effects mentioned above. This is another case of rigging-up the FFS; it is not a case of top-down effects on early visual processing.

The early effects may also be explained by invoking configurations of properties of objects or scenes. Neurophysiological research (Grill-Spector et al. 2006), psychological research (Peterson and Enns 2005), and computation modeling (Ullman et al. 2002) suggest that what is stored in early visual areas are implicit associations representing fragments of objects and shapes ("edge complexes"), as opposed to whole objects and shapes. One of the reasons that researchers hold that it is object and shape fragments that are used in rapid classifications instead of hole objects and shapes is the following; If these associations affect figure-ground segmentation, in view of the fact that figure-ground segmentation occurs very early (80–100 ms) (Lamme and Roelfsema 2000), they must be stored in early visual areas (up to V4, LO and posterior IT); early visual areas store object and shape fragments that speed up the FFS.

These associations reflect the statistical distribution of properties in environmental scenes (Delmore et al. 2004). The statistical differences in physical properties of different subsets of images are detected very early by the visual system before any top-down semantic involvement, as is evidenced by the elicitation of an early deflection in the differential between animal-target and non-target ERP's at about 98 ms in the occipital lobe. The low-cues are retrieved by analyzing the energy distribution across a set of orientation and spatial frequency tuned channels (Torralba and Oliva 2003).

As I have argued, the early latency of the effects of perceptual learning precludes the possibility of this being a top-down, cognitive effect. It is, instead, a bottom-up, stimulus-driven effect. Nevertheless, one could maintain that the information stored in visual circuits as a result of perceptual learning involves sensory concepts rendering early vision conceptual and, this, CP from the within. I presented evidence suggesting that the early classification is due to associations of shape and object fragments stored in early visual areas. I have explained elsewhere (Raftopoulos 2009) why these associations do not function the way concepts do and, thus, cannot be construed as concepts in a philosophically interesting way. Let me say here only that these associations can hardly be sensory *concepts* because they are implicit and can be used only for one purpose. Concepts, on the other hand, are explicit in principle and can be used in a variety of contexts.

3.2 Extrinsic Cognitive Effects on Early Vision

I claimed above that there are no intrinsic cognitive effects affecting early vision either in a top-down manner, or from within. Things differ, however, with respect to the extrinsic cognitive effects on early vision, because such effects clearly are found. Consider, for example, the role of spatial attention as the determinant of the locus of focus. The choice of this locus depends on some cognitive factors and this choice clearly co-determines the phenomenal content. It follows that despite the fact that this effect of spatial attention is not intrinsic to early vision, one might justifiably claim that since a causal explanation of why the perceiver is in a state with a specific content involves the cognitive contents of the states guiding attention, cognition affects early vision rendering it CP. Tye (1995, p. 140) thinks that the differences in the phenomenal content of perception when seeing ambiguous figures, such as the duck/rabbit figure as a duck-like or as a rabbit-like, may be causally influenced by the conceptual abilities of the viewer, which, thus, affect the phenomenal content of experience. Since the role of spatial attention in decomposing, organizing a visual scene, and determining the way ambiguous figures are perceived is well established, why not say that early vision is CP because a causal explanation of the phenomenal content involves concepts? That concepts driving spatial attention enter in an explanation of the percept renders perceptual processes CP. To recapitulate, even if one manages to show that early vision is not intrinsically affected by cognition, early vision is affected by cognition indirectly or extrinsically. Thus, one has to decide whether the extrinsic effects on early vision entail the CP of early vision.

As I claimed in Sect. 1, in determining whether extrinsic cognitive effects on perception should count as instances of CP, one should consider whether these effects vitiate the justificatory role of perception, and in this case, of early vision, in grounding perceptual beliefs. There are two sorts of indirect attentional effects and one has to consider the epistemological criterion to determine whether these effects should count as CP. The first is the role of spatial attention in determining the locus of focus. The second is the pre-cueing effects in perception.

3.2.1 Spatial Attention as Determinant of the Locus of Focus

Let us apply the epistemological criterion for judging whether the indirect conceptual influences on early vision through the determination of the spatial focus should be considered as instances of CP. The following discussion is meant to provide only a sketch of the springboard on which an in-depth discussion of the epistemological effects of cognitive penetration of perception should be based and is not meant in any way as an answer to the problem. Suppose that X and Y view the duck/rabbit ambiguous figure. X, by activating for some reason or other, the concept 'rabbit', decomposes and organizes the figure in such a way that X sees a rabbit. Y, on the other hand, by activating the concept 'duck' organizes the figure in a different way, and, as a result, sees a duck. If the role of spatial attention were to constitute a case of genuine CP, it should preclude early vision from playing the epistemological role of providing a theory neutral basis on which to resolve matters pertaining to seeing. Let us, thus, examine whether early vision allows the resolution of differences related to seeing despite the cognitive influences on perception through spatial attention. Suppose that cognitive factors have given rise to a context in which *X* is biased towards rabbits and hence, expects a rabbit, because the concept 'rabbit' is activated and used. When *X* is presented with a rabbit/duck ambiguous shape, focusing her attention onto the location at which she expects the characteristic ears, which in the standard drawings of the ambiguous figure is the upper part of the image, sees a rabbit. Y, who expects a duck because of the activation of the concept 'duck' owing to a different set of cognitive factors, focuses onto the lower part of the picture and sees a duck.

This holds true because it is well documented by a host of empirical studies with bi-stable stimuli (Attneave 1971; Britz and Pitts 2011; Driver and Baylis 1996; Hochberg and Peterson 1987; Kornmeier et al. 2011; Peterson and Hochberg 1983; Pitts et al. 2007) that shed light on the mechanisms that underlie the way perceptual set biases object segmentation, that the cognitive states of the observer do not affect by themselves the organization of the stimulus. Some crucial points of fixation influence the organization of the stimulus through the role of spatial attention; that is, there are locations in the image fixation on which favors one or the other percept. In other words, the way a bi-stable stimulus can be visually interpreted depends on where the observer fixes her attention, because there are crucial points fixation on which determines the perceptual interpretation. This means that the mechanism underlying the effect of perceptual set in ambiguous figures involves the voluntary control of spatial attention; the perceptual set induces observers to allocate their attention to specific regions in the stimulus (Peterson and Gibson 1994). This causes the figure to be experienced the way favored by perceptual set.

Let us assume now that X is instructed to lock her attention onto the lower part of the picture, instead onto the upper part at which her cognitive stances had led her to focus in the first place. Under these circumstances she would see a duck-like figure, as does Y. Note that X may see the duck-like figure, that is a configuration that a person who possesses the concept 'duck' would recognize as a duck, even when X does not possess the concept 'duck'. The important thing is that perceivers can shift attention despite their differing theoretical commitments since spatial attention can be controlled. Thus, a situation arises in which two persons with differing perceptual biases have their early vision outputting the same representation, a duck-like shape. Given this, and on the quite plausible assumption that they share a basic vocabulary, they could communicate with each other and understand what the other sees even though they initially saw different things.

The point is that although in the case of ambiguous figures cognition mediates the processes determining the percept through the allocation of spatial attention, and, thus the contents of the relevant cognitive states will enter in a causal explanation of why this specific percept was formed, spatial attention can be controlled since people can be instructed to focus their attention on such and such a location despite the fact that they may have entirely different intentional states. This way, the cognitive influences could be mitigated. Given that spatial attention is the factor that causes the formation of the two different percepts, controlling for this factor dissipates the difference and, thus, theoretical differences do not to affect the course of information retrieval from a scene. Once focusing at a certain location or part of the image takes place, the information is retrieved from the visual scene in a bottom-up way and is not affected by any concepts. This entails that the same stimulus would produce the same percept if the focus were the same because cognition through spatial attention guides the choice of the sites of focusing but does not affect the perceptual process (Raftopoulos 2009).

As I have said, most definitions of CP do not explain why such extrinsic cognitive effects on early vision should not count as CP. A usual justification of this claim is that spatial attention acts externally to perception by determining the point of focus but does not affect perceptual processing itself and CP is supposed to be a matter of an internal, mental link between cognitive and perceptual states (Stokes 2012). Although this last demand on CP is correct, the problem with this explanation is that attention does not affect perception solely by determining the input externally to the perceptual system and by leaving unaffected perceptual processing; it also affects it. In late vision, attention affects processing intrinsically. Spatial attention also affects intrinsically early vision, except that the effects are data-driven and not cognitive. Thus, it is not enough to say that spatial attention does not entail CP because it acts only externally to determine the locus of attention. Even in early vision where there are no intrinsic *cognitive* effects of attention, and, thus, attention extrinsically only affects the processes of early vision without affecting the processing itself in an online way, still the extrinsic effects modulate indirectly the internal mental on-goings by setting the parameters of the equations that transform states.

I have attempted to provide an adequate explanation why extrinsic cognitive effects do not constitute instances of CP, by introducing the epistemological condition for CP and by showing that these effects do not undermine the epistemological role of perception. The matter is not closed, however, as there is a second sort of indirect/extrinsic cognitive attentional effects on early vision that may entail that early vision is CP, namely, the pre-cueing effects.

3.2.2 Pre-cueing Effects in Perception

Attention may affect all stages of visual processing depending on the experimental set up. When a viewer observes a scene, sustained attention, whether it be spatial or object/feature based, intervenes late enough so that it does not affect early vision and, thus, these effects do not entail the CP of early vision. Note that cognitively-driven attention does not entail the CP of early vision not because it acts externally to perception, but because it affects a later stage of visual processing, and in this sense it is extrinsic to early vision. In contradistinction, attentional effects are intrinsic in late vision, rendering it CP.

When a viewer has been instructed to attend to a certain location or to expect a certain feature or object to appear, attention affects perception by modulating the internal on-goings biasing the base-line activation of the neurons that encode the expected stimulus or location. By being internal and not external, this short of attentional effects is a viable candidate as a cause of CP of early vision.

Studies of the effects of spatial attention cues presented to a viewer before stimulus presentation show early modulation of perceptual processing (Carrasco 2011; Reynolds and Chelazzi 2004). This phenomenon refers to the enhancement of the baseline activity of neurons at all levels in the visual cortex that are tuned to a location that is cued and thus the location attracts attention before the onset of any stimuli. It is called *attentional modulation of spontaneous activity*. The spontaneous firing rates of neurons are increased when attention is shifted toward the location of an upcoming stimulus before its presentation.

This cueing is thought to reflect the effects of the neural processes that occur in response to instructions or cues to orient attention to a specific location before the stimulus appears. Spatial attention enhances the sensitivity of the neurons tuned to the attended spatial location by improving the signal-to-noise ratio of the neurons tuned to the attended location over the neurons with receptive fields outside the attended location that contribute only noise. This effect does not determine what subjects perceive in that location because by enhancing the responses of all neurons tuned to the attended location independent of the neurons' preferred stimuli it keeps the differential responses of the neurons' unaltered and thus does not affect what it is perceived. To put it differently, spatial attention determines the focus of the gaze but does not solve the gazing problem of attention. What is perceived depends on the relative activity of appropriate assembles of neurons that selectively code the features of the stimulus compared to the activity of assemblies that do not code the features of the stimulus and thus contribute noise. Since the percept depends on the differential response of these assemblies, this effect of spatial attention by not evoking differential responses leaves the percept unchanged; it makes detection of the objects/features in the scene easier but it does not determine what the observer perceives.

Evidence (Carrasco 2011; Shibata et al. 2008) also suggests that through pre-cueing of object features (instructing a subject to look at a screen for a red object, for example) feature-based attention modulates prestimulus activity in the visual cortex. In fMRI experiments designed to examine the effects of feature attention to color and motion on the visual, frontal, and parietal areas, a cue appeared 1 s before the stimulus. The activity within the color sensitive visual areas and the motor sensitive visual areas was increased by attention to color and motion, respectively. The effects of pre-stimulus feature attention may act either as a preparatory activity to enhance the stimulus-evoked potentials within feature sensitive areas, or they may act to modulate stimulus-locked transients.

Both effects of pre-cueing reflect a change in background neural activity. These biases constitute a case of rigging-up the feedforward sweep (FFS) in visual processing. They are anticipatory, occur before the stimulus presentation, and do not emerge as part of perceptual competition and in this sense they are not intrinsic to the perceptual processing (Nobre et al. 2012, p. 161), which is otherwise unaffected by top-down effects. In other words, what they do is to set up the values of some parameters that will affect the FFS processing. During the FFS there is no top-down cognitive activity to modulate the perceptual processing, which is data-driven. This means that the resulting biasing can be controlled for the same reasons that spatial attention focusing can be controlled and, thus, that cognitive effects of this sort could be mitigated. Two perceivers who might perceive different percepts given the same stimulus on account of some sort of pre-cueing may 'interchange' percepts by receiving one the cues of the other. Since the perceptual processing is otherwise not affected, controlling for the cues controls the percept. This negates the harmful effects of the role of concepts and, hence, even though attention affects the internal perceptual on-goings, these effects do not count as a case of CP.

4 Concluding Discussion

I have argued for two adequacy conditions that a definition of CP/CI should satisfy. The first condition is that any direct/intrinsic effects on perception or a stage of it entail CP. The second condition dictates that indirect/extrinsic effects on perception or a stage of it should count as cases of genuine CP if they undermine the role of perception or a stage of it as a neutral, concept free, arbiter of perceptual beliefs, and scientific theories. Applying these two criteria shows that early vision is CI, whereas late vision is CP.

References

Attneave, F. (1971). Multistability in perception. *Scientific American, 225*, 63–71.

Britz, J., & Pitts, M. (2011). Perceptual reversals during binocular rivalry: ERP components and their concomitant source differences. *Psychophysiology, 48*, 1489–1498.

Burge, T. (2010). *Origins of objectivity*. Oxford: Clarendon Press.

Carrasco, M. (2011). Visual attention: The past 25 years. *Vision Research, 51*, 1484–1525.

Chaumon, M., Drouet, V., & Tallon-Baudry, C. (2008). Unconscious associative memory affects visual processing before 100 ms. *Journal of Vision, 8*(3), 1–10.

Churchland, P. M. (1988). Perceptual plasticity and theoretical neutrality: A reply to Jerry Fodor. *Philosophy of Science, 55*, 167–187.

Crouzet, S. M., Kirchner, H., & Thorpe, S. J. (2010). Fast saccades toward faces: Face detection in just 100 ms. *Journal of Vision, 10*(4), 1–17.

Davis, M. (1995). Tacit knowledge and subdoxastic states. In C. MacDonald & G. Macdonald (Eds.), *Philosophy of psychology: Debates on psychological explanation*. Oxford: Blacwell.

Delmore, A., Rousselet, G. A., Mace, M. J.-M., & Fabre-Thorpe, M. (2004). Interaction of top-down and bottom up processing in the fast visual analysis of natural scenes. *Cognitive Brain Research, 19*, 103–113.

Dennett, D. C. (1983). Styles of mental representation. *Proceedings of the Aristotelian Society, 83*, 213–226.

Driver, J., & Baylis, G. S. (1996). Eye-assignment and figure-ground segregation in short-term visual matching. *Cognitive Psychology, 31*, 248–306.

Fodor, J. (1983). *The modularity of mind*. Cambridge, MA: The MIT Press.

Grill-Spector, K., Kushnir, T., Hendler, T., Edelman, S., Itzchak, Y., & Malach, R. (1998). A sequence of object-processing stages revealed by fMRI in the Human occipital lobe. *Human Brain Mapping, 6*, 316–328.

Grill-Spector, K., Henson, R., & Martin, A. (2006). Repetition and the brain: Neural models of stimulus-specific effects. *Trends in Cognitive Sciences, 10*, 14–23.

Hanson, N. R. (1958). *Patterns of discovery*. Cambridge: Cambridge University Press.

Haugeland, J. (1998). *Having thought*. Cambridge: Harvard University Press.

Hochberg, J., & Peterson, M. A. (1987). Piecemeal organization and cognitive components in object perception. *Journal of Experimental Psychology: General, 116*, 370–380.

Johnson, J. S., & Olshausen, B. A. (2005). The earliest EEG signatures of object recognition in a cued-target task are postesensory. *Journal of Vision, 5*, 299–312.

Kirchner, H., & Thorpe, S. J. (2006). Ultra-rapid object detection with saccadic movements: Visual processing speed revisited. *Vision Research, 46*, 1762–1776.

Kornmeier, J., Pfaffle, M., & Bach, M. (2011). Necker-cube: Stimulus-related (low-level) and percept-related (high-level) EEG signatures early in occipital cortex. *Journal of Vision, 11*(9), 1–11.

Kuhn, T. S. (1962). *The structure of scientific revolutions*. Chicago: Chicago University Press.

Lamme, V. A. F., & Roelfsema, P. R. (2000). The distinct modes of vision offered by feedforward and recurrent processing. *Trends in Neuroscience, 23*, 571–579.

Liu, H., Agam, Y., Madsen, J., & Krelman, G. (2009). Timing, timing, timing: Fast decoding of object information from intracranial field potentials in human visual cortex. *Neuron, 62*, 281–290.

Macpherson, F. (2012). Cognitive penetration of colour experience. *Philosophy and Phenomenological Research, 84*(1), 24–62.

Marr, D. (1982). *Vision: A computational investigation into human representation and processing of visual information*. San Francisco: Freeman.

Nobre, A. C., Rohenkhol, G., & Stokes, M. G. (2012). Nervous anticipation: Top-down biasing across space and time. In M. Posner (Ed.), *Cognitive neuroscience of attention* (2nd ed.). New York, N.Y: The Guilford Press.

Peterson, M. A., & Hochberg, J. (1983). Opposed set-measurements procedure: A quantitative analysis of the role of local cues and intention in form perception. *Journal of Experimental Psychology: Human Perception and Performance*, *9*, 183–193.
Peterson, M. A., & Gibson, B. S. (1994). Object recognition contributions to Figure-ground organization: Operations and outlines and subjective contours. *Psychological Science*, *5*, 253–259.
Peterson, M. A., & Enns, J. (2005). The edge complex: Implicit memory for figure assignment in shape perception. *Perception and Psychophysics*, *67*(4), 727–740.
Pitts, M., Nerger, J., & Davis, T. J. R. (2007). Electrophysiological correlates of perceptual reversals for three different types of multistable images. *Journal of Vision*, *7*(1), 1–14.
Pylyshyn, Z. (1999). Is vision continuous with cognition? *Behavioral and Brain Sciences*, *22*, 341–365.
Pylyshyn, Z. (2003). *Seeing and visualizing: It's not what you think*. Cambridge, MA: The MIT press.
Raftopoulos, A. (2001a). Is perception informationally encapsulated? The Issue of the Theory-Ladenness of Perception. *Cognitive Science, 25*, 423–451.
Raftopoulos, A. (2001b). Reentrant pathways and the theory-ladenness of observation. *Philosophy of Science*, *68*, 187–200.
Raftopoulos, A. (2006). Defending realism on the proper ground. *Philosophical Psychology*, *19*(1), 1–31.
Raftopoulos, A. (2009). *Cognition and Perception*. Cambridge, MA: The MIT Press.
Raftopoulos, A. (2013). The cognitive impenetrability of the content of early vision is a necessary and sufficient condition for purely nonconceptual content. *Philosophical Psychology*, 1–20. doi:10.1080/09515089.2012.729486.
Reynolds, J. H., & Chelazzi, L. (2004). Attentional modulation of visual processing. *Annual Review of Neuroscience*, *27*, 611–47.
Searle, J. R. (1995). Consciousness, explanatory inversion and cognitive science. In C. MacDonald & G. Macdonald (Eds.), *Philosophy of psychology: Debates on psychological explanation*. Oxford: Blacwell.
Sellars, W. (1956). Empiricism and the philosophy of mind. In H. Feigl & M. Scriven (Eds.), *Minnesota studies in the philosophy of science* (Vol. I). Minneapolis, MN: University of Minnesota Press.
Shibata, K., Yamagishi, N., Naokazu, G., Yoshioka, T., Yamashita, O., Sato, M., et al. (2008). The effects of feature attention on prestimulus cortical activity in the human visual system. *Cerebral Cortex*, *18*, 1644–1675.
Siegel, S. (2012). Cognitive penetrability and perceptual justification. *Nous*, *46*, 201–222.
Spelke, E. S. (1988). Object perception, In A. I. Goldman (Ed.), *Readings in philosophy and cognitive science* (pp. 447–461). Cambridge, MA: MIT Press.
Stich, S. (1978). Beliefs and subdoxastic states. *Philosophy of Science*, *45*, 499–518.
Stokes, D. (2012). Perceiving and desiring: A new look at the cognitive penetrability of experience. *Philosophical Studies, 158*(3), 479–92.
Torralba, A., & Oliva, A. (2003). Statistics of natural image categories. *Network*, *14*, 391–412.
Tye, M. (1995). *Ten problems of consciousness*. Cambridge, MA: The MIT Press.
Ullman, S., Vidal-Naquet, M., & Sali, E. (2002). Visual features of intermediate complexity and their use in classification. *Nature Neuroscience*, *5*(7), 682–687.

The Emergence and Development of Causal Representations

Xiang Chen

Abstract In this article I intend to analyze how infants' initial causal representations emerge and how their primitive causal knowledge evolves during cognitive development. Recent studies from developmental psychology report that infants acquire knowledge of physical causality earlier than what Piaget has described—by 6 months of age infants have already shown signs of sensibility to causality in simply physical events such as mechanical collision. Studies also suggest that infants recognize causality at such an early age by activating a primitive schema that uses merely spatial and temporal cues to identify causal relations. As infants grow, they learn to use complex information to identify causal relations, and they eventually acquire sophisticated schemas to represent causality. However, the development of causal knowledge is not a linear progression as Piaget has suggested. Infants' primitive causal schema never disappears completely when infant grow and acquire mature understanding of causal relations. Infants' primitive causal schema continues to exist in adulthood and plays a role in adults' cognitive processing. When adults' cognitive system is overloaded by processing information required by sophisticated causal schemas, it would protect itself by falling back to a lower level of information processing and returning to the primitive causal schema. When adults fall back to the primitive causal schema and overextend it beyond the realm of mechanical causation, misconceptions are bound to occur.

1 Introduction

The moment when causal representation are formed is a milestone in cognitive development. According to Piaget, the development of causal representations is a result of biological maturation and experiential learning, and it can be divided into several discrete stages with qualitative differences (Piaget 1954). For example, around 6

X. Chen (✉)
Department of Philosophy, California Lutheran University,
Thousand Oaks, CA, USA
e-mail: chenxi@clunet.edu

L. Magnani et al. (eds.), *Philosophy and Cognitive Science II*,
Studies in Applied Philosophy, Epistemology and Rational Ethics 20,
DOI 10.1007/978-3-319-18479-1_2

months of age, infants begin to have a sense of causality through their secondary circular reactions, that is, their repeated actions that produce effects in the environment. Around 12 months of age, infants gradually recognize the existence of external causal relations after they learn to distinguish between themselves and their environment and to understand the permanence of objects. By the second year infants are able to distinguish between psychological causality as power over one's actions and physical causality as relationships between external objects. To Piaget, these stages form a progressive series—later stages contain increasingly sophisticated and complex schemas which help infants to comprehend causal relations. As infants grow, simplistic and immature schemas adopted in an earlier stage are abandoned and replaced by sophisticated and complex ones in a later stage. After infants acquire mature understanding of causal relations, there is no reason to return to previous immature knowledge. For the same reason, there is no reason for adults to retune to infants' immature world. The development of causal knowledge is a linear process as we grow.

How do infants recognize causality? Given infants' limited cognitive capacities, it is reasonable to expect that they probably recognize causality and construct causal representations by means of mechanisms with unique characteristics dissimilar to those underneath adults' causal representations. If so, what are the cognitive mechanisms responsible for infants' causal representations and how are they different from adults'? Furthermore, how do infants' initial causal representations evolve during cognitive development? After infants grow and become mature, are the cognitive mechanisms responsible for their initial causal representations overturned and replaced by another set of cognitive mechanisms as what Piaget suggested, or do they continue to exist and play a role in adults' cognitive processing?

By considering the most recent studies from both developmental and cognitive psychology, I intend to review in this article how infants' initial causal representations emerge and how their primitive causal knowledge evolves during cognitive development. Recent studies from developmental psychology report that infants develop understanding of physical causality earlier than what Piaget had described—by 6 months of age infants have already shown sign of sensibility to causality in simply physical events such as mechanical collision. Studies also suggest that infants recognize causality at such an early age by activating a primitive schema that uses merely spatial and temporal cues to identify causal relations. As infants grow, they learn to use other kinds of information that is more complex than spatial and temporal cues to identify causal relations, and they eventually acquire sophisticated schemas to represent causality. However, the development of causal knowledge is not a linear progression as what Piaget had suggested. Infants' primitive causal schema never disappears completely when infant grow and acquire mature understanding of causal relations. Infants' primitive causal schema continues to exist in adulthood and plays a role in adults' cognitive processing. When adults' cognitive system is overloaded by processing information required by sophisticated causal schemas, it would protect itself by falling back to a primitive level of information processing and returning to the primitive causal schema. When adults fall back to the primitive causal schema and overextend it beyond the realm of mechanical causation, misconceptions are bound to occur.

2 The Emergence of Causal Representations

Over the last three decades, studies from developmental psychology had offered rich evidence that infants' understanding of physical events is more sophisticated than what Piaget had described. By 6 months of age infants have already shown sign of sensitivity to causality by distinguishing causal from non-causal relations embedded in simply physical events such as collision. When infants at this age are shown an event in which one object collides with another one and the second object moves immediately after the collision, they conclude, as adults would do, that the first object causes the second object to move (Leslie and Keeble 1987; Kotovsky and Baillargeon 2000; Chi et al. 2012).

It is difficult to investigate how preverbal infants think. Almost all evidence on infants' causal representations is indirect, coming from habituation experiments that employ the so-called method of looking time. The assumption of this method is that the more interesting infants find an event the longer they will watch it. A typical habituation experiment begins by showing the subjects (infants) an event several times until they lose their interest and spend less time looking at it. Then the subjects receive test events that are different from the one that they have just seen. An increase in looking time for a test event indicates that the subjects recognize its novel features. In this way, the method of looking time gives researchers a valuable insight in the mind of preverbal infants.

In a series of habituation experiments, Leslie in the 1980s found that, infants as young as 6 months of age were able to use spatial and temporal continuities as the criteria to identify causality. He began by habituating one group of infants with a direct launching sequence—one object collides with another object, which moves immediately after the collision. This is a typical causal event according to adults' experience. After habituation, Leslie showed the infants a launching sequence in which the first object stopped short of the second and after a short delay the second object moved off without being stuck. This is not a causal event according to adults' experience because of the spatial gap between the two objects and the temporal gap between the moments of impact and reaction. Like adults, infants in Leslie's study recognized the categorical differences between the causal and the non-causal events, indicated by increasing looking time to the event with spatial and temporal gaps. In comparison, the other group of infants saw a delayed sequence (a temporal gap between impact and reaction) in the habituation trial but a gapped sequence (a spatial gap between the two objects) in the test trial, and they showed little surprise. They probably recognized the similarities between the gapped and the delayed sequences—both are non-causal events (Leslie 1984).

Sensitivity to spatiotemporal continuity is not the same as the ability to recognize causality. To verify that young infants indeed can recognize causal relations, Leslie and Keeble designed a new procedure (Leslie and Keeble 1987). They trained 6-month-old infants with either a direct launching sequence or a delayed sequence. After habituation, they showed the infants sequences to which they had just habituated in reverse; that is, instead of moving from the left to the right, objects move from the right to the left. Reversal of a delayed launching sequence generates changes of spatiotemporal features (left/right orientation and order of movement). Contrary, reversal of a direct launching sequence produces not only changes of the spatiotemporal features but also changes of causal roles—the agent of the original sequence becomes the patient of the new sequence and the patient becomes the agent. Therefore, if infants are able to perceive causal relations, they should detect the changes of causal roles in the direct launching events, and consequently spend more time looking at the reversed direct launching sequence than the reversed delayed sequence. This was exactly how the infants behaved in Leslie and Keeble's study—they were indeed looked longer at the reversed direct launching sequence than the reversed delayed sequence.

Since infants can form causal representations at such an early age, Leslie reasoned that a fairly low-level perceptual mechanism such as visual processing, which presumably takes input from processes of motion perception, is responsible for infants' early causal perceptions. Leslie proposed a causal perception hypothesis to account for the cognitive mechanisms responsible for infants' causal recognition. This hypothesis assumes that there is a perceptual module that identifies causal relations by processing information about what have taken place in infants' visual environment (Leslie and Keeble 1987). Leslie's hypothesis has two important implications. First, like other low-level perceptual mechanisms, visual processing occurs in a fixed, automatic way without being influenced by information or reasoning abilities outside the module. The perceptual module responsible for infants' causal representations is encapsulated, using information only from perception and from the module itself. Second, because the perceptual module responsible for infants' causal representations is encapsulated, infants' ability to represent causality is neither the result of learning nor of the process of gradual development. Infant's ability to represent causality must be innate.

These implications from the causal perception hypothesis are controversial. First, if the ability to represent causal relations is hardwired for evolutionary reasons, it is reasonable to expect that infants should be able to distinguish causal from non-causal events early on, before 6 months of age. However, studies found that before 6 months of age infants show no sign of sensitivity toward causal relations. In a series of studies, Cohen and Amsel observed 4-month-old infants' responses to launching events. The infants were first habituated to a direct, a gapped, or a delayed launching sequence, and then they were tested with event sequences that differed from the habituation sequence either in terms of causality or in terms of individual temporal or spatial features. If these infants are sensitive to causal relations, they should look at a causal sequence longer when they were habituated with a non-causal one, or they should lost their interests quickly to a causal sequence when they were habituated with a causal one. The results however showed that 4-month-old infants in the studies always looked longer at causal events during the test phase regardless of whether they were habituated to causal or non-causal events. The infants in the studies continued to be interested as long as test events provided continuous movement, and they looked away if any interruption occurred such as a temporal delay or a spatial gap. These results suggested that these 4-month-old infants showed no sign of the sensitivity to causality that 6-month-olds' seem to have (Cohen and Amsel 1998). Similar studies were also conducted by Desrochers, who reported that 3.5-month-old infants did not perceive direct launching sequences as causal (Desrochers 1999). These findings challenge the notion that the ability to perceive causality is innate and hard-wired.

Second, if the perceptual module responsible for infants' causal representations is encapsulated, infants can't develop the ability to represent causality through learning. However, studies find that infants can recognize causality before 6 months of age if they are given real-world experience of causal action (Rakison and Krogh 2012). Prior to habituation, Rakison and Krogh had a group of 4.5-month-old infants wore mittens covered in Velcro and allowed them to interact with balls that were also covered in Velcro. With the help of Velcro, these infants were able to play with the balls. Rakison and Krogh had another group of infants wore non-sticky mittens and gave them non-sticky balls to play. Without Velcro, these infants were not able to interact with the ball. Then both groups were habituated and tested according a procedure similar to the one used by Leslie and Keeble in their 1987 studies. If real-world causal action does not facilitate causal perception, wearing sticky mittens and interacting with sticky balls should not affect the performance of the subjects—4.5-month-old infants, as reported by Cohen and Amsel, are not able to recognize causal relations. Surprisingly, 4.5-month-old infants who had interacted with objects by wearing sticky Velcro mittens in the study seemed to interpret direct launching sequences in terms of causality. Following habituation with a direct launching event, they showed strong interests to the event in which the agent and recipient relation was reversed. In contrast, infants who could not interact with the balls because they wore non-sticky mittens did not distinguish the differences generated by the reversal of the launching direction. Rakison and Krogh's study provides strong evidence that infants younger than 6 months of age are capable of perceiving causality in simple launching events through brief learning experience. Contrary to Leslie's claim, the ability to represent causal relations can be affected by outside influence.

Recently, Rips proposes a different hypothesis to account for the cognitive mechanisms behind infants' causal representations (Rips 2011). Instead of appealing to a hardwired, encapsulated perceptual mechanism, Rips hypothesizes that infants recognize and represent causal sequences through activating a schema stored in long-term memory. A schema is a complex structure of knowledge, acquired and improved through learning from experience. Rips assumes that there is causal schema in long-term memory that provides infants with a general description of causal relations embedded in a direct launching sequence. For example, this schema specifies that each launching sequence has an agent and a patient. It also specifies the routine of the causal event: it begins with an approach motion of the agent, then an impact between the agent and the patient, and immediately a withdrawal motion of the patient.

As an infant encounters a launching sequence, the initial observation would activate the causal schema, which assigns the agent and patient roles to the sequence and activates expectations about what will occur next. If the schema's expectations are confirmed, the infant would conclude that a causal interaction has taken place in the launching sequence.

This causal schema uses only spatiotemporal parameters to identify causal relations embedded in a launching sequence. All of the key features of a causal event, including the approach of the agent, the impact between the agent and the patient, and the immediate withdrawal of the patient, can be effectively defined by means of spatial and temporal cues. Spatial and temporal continuity thus becomes the criterion to identify causality. In this way, infants' causal schema highlights two important characters of causality. First, because an impact between the agent and the patient is necessary, causation is direct in the sense that an object is causally related to another object if and only if they touch (causation on contact), and that there are no causal relations between two objects if they do not come into contact (no causation at a distance). Second, because the withdrawal of the patient occurs immediately after the impact, causation is instant in the sense that an object is causally related to another object if and only if the reaction of the second objects occurs immediately after the contact and there is no time delay between the impact and the reaction.

3 The Development of Causal Representations

Studies from developmental psychology indicate that infants begin to understand many important features of objects at a very early age. 2.5-month-old infants already know that it is possible to insert an object into a container with an open top but not into a container with a closed top, and that an object inside a container cannot pass through its wall and hence must remain in it, and move with it, until removed through its opening (Aguiar and Baillargeon 1999; Hespos and Baillargeon 2001). By four months of age, infants are clearly aware of object solidity, that is, two objects cannot occupy the same space at the same time (Baillargeon et al. 1985; Spelke et al. 1992). This kind of early comprehension of object solidity is a part of our "core cognition," a system of representations created by innate perceptual devices (Spelke and Kinzler 2007; Carey 2009). As a result of natural selection, we are born with several perceptual analyzers, which allow us to individualize and identify certain kinds of ontological entities such as objects at an early age.

Studies further show that infants in the early stage of their lives use only a specific kind of information in the task of object individualization. In theory, an object can be individualized by means of spatiotemporal information (location and movement), property information (color, shape, size, texture and so on), or kind information (category membership). It is found that infants as young as 4 months of age can use spatiotemporal information to distinguish objects from each other (Spelke and Kestenbaum 1986), but they are not able to use property and kind information to individualize objects until a much later age. In a series of studies, Xu and Carey discovered that even at 10 months of age infants were still not able to individualize objects with property information. They first showed 10-month-old infants a screen with two objects with different colors, different textures and different shapes (e.g. a yellow rubber toy duck and a white foam ball) concealed behind. Next, one object (say, a yellow rubber toy duck) emerged from behind the screen to its left and returned behind the screen, and then another object (a white foam ball) emerged from behind the screen to its right and returned behind the screen. Each object emerged four times in this experiment. Because the screen occluded the movements of the objects and because the two objects never appeared simultaneously, this setting did not provide the infants with any direct spatiotemporal information. To correctly determine how many objects were behind the screen, the infants must rely upon property information. In these studies, 10-month-old infants failed to determine correctly that there were two objects behind the screen when they were offered only property information. However, the same subjects succeeded to infer the correct number of objects behind the screen when they were provided with additional spatiotemporal information: before their alternate movements began, the two objects were brought out from behind the screen simultaneously, one to each side of the screen, for about 3 s. Xu and Carey concluded that, infants under 10 months of age prefer to use spatiotemporal over property information in tasks of object individuation, and that not until 12 months of age can infants use property and kind information to do so (Xu and Carey 1996; Xu 1999).

The privilege of spatiotemporal information in infants' core object cognition probably comes from its simplicity. Spatiotemporal information is simple in the sense that it can be employed to individuate objects without any background or prerequisite information—spatiotemporal discontinuity alone warrants an inference of separate objects. In object individuation tasks, spatiotemporal information is used to form mental tokens that function as pointer to indicate objects' locations. Because object indexing by means of location requires only a minimal level of cognitive processing, infants are able to use spatiotemporal information to individualize objects at a very early age (Leslie et al. 1998).

When infants grow, however, spatiotemporal information is no longer privileged. Complex and sophisticated information frequently overrides spatiotemporal information in tasks of object individuation. For example, adults know that a living being ceases to exist and turns into another kind when it dies, in spite of the spatiotemporal continuity of its body—an example of kind information overrides spatiotemporal information. According to Cohen and his colleagues (Cohen and Cashon 2001; Cohen et al. 2002), moving from processing simple and primitive information to complex and sophisticated information is a hallmark of progress in cognitive development. In the development of object knowledge, infants first use simple spatial and temporal cues to identify and individualize objects. Spatiotemporal information alone, however, can't always warrants individualization of objects, as shown by Xu and Carey's studies. In the next step of the cognitive development, infants are able to use property information to identify and individualize objects. Property information is richer and more complex than spatial and temporal cues, and thus requires a higher level of cognitive processing. But property information alone cannot always warrant distinctions of different objects either. To know whether a radical difference in size warrants existence of different objects, for example, we need information regarding their categories. A radical size difference warrants the inference of two distinct chairs, but not two distinct plants (Xu and Carey 1996). In another step of the cognitive development, infants acquire the ability to use kind information to identify objects. Information about the category of objects is at an even higher level of complexity, which requires using distinct labels to refer to distinct types of objects, a capacity that develops through word learning (Xu 1999). As infants grow, they develop the ability to process complex information by integrating partial, incoherent, or discrete information into completed, coherent and holistic unit. In this way, cognitive development involves progress from processing relatively simple information to processing complex and sophisticated information.

We can also see the patterns of moving from simple to complex, from partial to completed, from incoherent to coherent, and from discrete to holistic information in the development of causal representations. The first sign of infants' causal sensitivity appears around 5.5 months of age (Cohen and Amsel 1998). In a study, Cohen and Amsel first habituated 5.5-month-old infants to a direct, a gapped, or a delayed launching sequence. In the test phase, Cohen and Amsel showed the subjects event sequences different from the habituation sequence either in terms of causality (e.g. habituated with a causal sequence and tested with a non-causal one) or in terms of individual temporal or spatial features (e.g. habituated with a gapped sequence and tested with a delayed sequence). They found that infants who were habituated to a causal sequence still looked longer at the causal sequence than either one of the non-causal ones, a sign that the infants did not recognize the difference in terms of causality between these two types of event. Cohen and Amsel also found that those infants habituated to a non-causal sequence were surprised when they were showed the other type of non-causal sequence in the test phase. This finding suggested that the infants might have analyzed these sequences in terms of their independent spatial and temporal features. Infants habituated to a gapped sequence, for example, saw a spatial gap and a temporally continuous motion during habituation. They were surprised during the test phase because they saw two perceptual changes in the delayed sequence—the spatial gap disappeared and the continuous motion became discontinuous.

A crucial breakthrough in the development of causal representations occurs around 6 months of age. Cohen and Amsel tested 6-month-old infants with the same procedure, and found that for the first time infants were able to distinguish causal from non-causal events: infants habituated to either type of non-causal events spent more time looking the causal event in the test phase than the other type of non-causal event, and those habituated to the causal event spent more time looking at the either type of non-causal events than the causal one (Cohen and Amsel 1998). To do so, the infants must have processed the spatial and temporal cues in ways different from how 5.5-month-old infants did. If spatial and temporal cues were treated independently as what 5.5-month-old infants did, there should be greater dissimilarities between two non-causal sequences than between a causal sequence and a non-causal one, and infants should spend more time looking at a non-causal than a causal event in the test phase if they habituated to the other type of non-causal event. Thus, to recognize causality in this setting, the 6-month-old infants in this study must have developed an ability to perceive a collision event as a whole in terms of the spatiotemporal relationships between the involved objects, rather than simply processing the spatiotemporal characteristics of each object individually.

Similar to the development of object individuation and representations, development in causal representations also involves progress from processing simple and discrete to complex and sophisticated information. As infants grow, they first develop the ability to process spatial and temporal cues of each object involved in launching sequences individually. At this stage, information is partial, inherent, and discrete. Next, they develop the ability to treat a launching sequence as a whole and to process spatial and temporal relationships between the objects in the sequences, which enable them to recognize causality. This is a progress in which partial, incoherent, or discrete information is integrated into completed, coherent and holistic unit.

4 Falling Back to the Primitive Causal Schema

In general, it is more effective and efficient to learn by processing information that is completed, coherent and holistic. Thus, we have a tendency to process information at the highest level available, and adults consequently prefer processing sophisticated information over primitive spatiotemporal information. However, this does not mean that processing partial, incoherent, or discrete information becomes irrelevant when adults have developed the ability to process sophisticated and complex information. Sometimes adults have to "fall back" to a lower level of information processing even when sophisticated and coherent information processing is available. Such a "falling-back" reaction typically happens when the cognitive system becomes overloaded by processing information at a sophisticated and complex level (Cohen and Cashon 2001). Many circumstances can cause the cognitive system to be overloaded, but most commonly overloaded accidents occur when the input information contains a high level of noise (conflicting or irrelevant information), or when the load of information processing is too high, such as when the input information is unfamiliar and needs to be classified. When these circumstances happen, typical reactions of the cognitive system are to adopt a self-protection measure, falling back to a lower level of information processing to avoid a system breakdown.

Oaks and Cohen offered an example of the falling-back reactions from the development of causal representations (Oakes and Cohen 1990). They found that 6-month-old infants' sensibility to causation was limited—they could recognize causal relations in launching events only when the objects involved in the events were simple geometrical shapes such as a red circle or a green square. When the objects involved in a launching sequence were realistic items such as toy vehicles, 6-month-olds failed to recognize causality. They apparently fell back to processing spatial and temporal cues of each object individually and did not perceive launching sequences as a whole. 10-month-old infants, however, could handle the extra load of information presented by complex objects and could perceive causality in realistic events correctly. But 10-month-old infants' ability to handle complex, realistic casual sequences was not stable either. When they encountered even more complicated situations, they too fell back to a lower level of information processing. Cohen and Oakes performed basically a replication of their own 1990 study on 10-month-old infants with only one difference. Instead of using a single toy vehicle as the objects, they showed the infants five different pairs of vehicles during habituation (Cohen and Oakes 1993). Cohen and Oakes found that 10-month-olds in this setting failed to recognize causality: infants habituated to a non-causal event dishabituated to other non-causal sequences rather than the causal one. These studies indicated that whether infants could recognize causality embedded in a launching sequence depended in part upon the load of information. When a launching sequence involved complex objects, infants who might have been able to perceive causality under simpler conditions could no longer do so. As a reaction to information overload, they fell back to a primitive level of information processing, analyzing the sequence merely in terms of independent spatial and temporal parameters, and they subsequently failed to recognize the embedded causality.

When infants fall back to a primitive level of information processing in the evolution of causal knowledge, they return to the primitive causal schema acquired in their early age and they inevitably overextend it into situations that require sophisticated information processing. Consequently, misconceptions are bound to happen. We can find an interesting example of misconceptions associated with falling back to the primitive causal schema in infants' understanding of shadows.

Shadows are not physical objects, and motions of shadows violate the primitive causal schema that describes causal relations between objects. When an object casts a shadow onto a surface, the shadow does not moves with the surface, but with the object that casts it. The relationships among the object, the shadow and the surface violate both the constraint of causation on contact and the constraint of no causation at a distance defined by the primitive causal schema. Spelke and her cooperators designed a series of studies to investigate whether infants could understand motions of shadows correctly and whether they could distinguish motions of shadows from motions of objects (Spelke et al. 1995). In their studies, Spelke and her cooperators first habituated 8-month-old infants with a stationary display consisting of a shadow, a ball that appeared to cast the shadow, and a box on which the shadow rested. The test trials involved two different scenarios. In the first scenario, the ball moved and the infants were showed two event sequences: either the shadow moved with the ball or remained at rest. The former is a natural motion but in violation of the constraint of no causation at a distance, and the latter is unnatural but consistent with the constraint. A majority of the infants in the studies were surprised by the natural event in which the motion of the shadow violated the primitive schema. These infants incorrectly inferred that the shadow should remain at rest probably because the shadows and the ball are spatially separated and they didn't anticipate causation at a distance. In the second scenario the box moved and the infants saw either the shadow remain at rest or moved with the box. The former is natural but in violation of the constraint of causation on contact, and the latter is unnatural but consistent with the constraint. Again, a majority of the infants in the studies were surprised by the natural event in which the shadow remained at rest. Infants appeared to infer that the shadow should move with the surface because they supposed causation on contact as defined by the primitive schema. Studies by Spelke and her cooperators suggested that these infants made false inferences about the motion of shadows by overextending the primitive causal schema into causal relations between different kinds of entities. This tendency to overextend knowledge of mechanical causation appears to persist well into later stages of cognitive development, generating numeral misconceptions in both children's and adults' understanding of the world (Van de Walle et al. 1998).

Misconceptions associated with the primitive causal schema also exist in real-life situations. We can find an example of these misconceptions in the learning of solid physics. Typically, introductory physics courses start with discussions of point particles. Students, both at the high school and the college levels, usually do not experience too much trouble in comprehending particle motions, because particle dynamics are consistent with the primitive causal schema and can be analyzed merely in terms of spatial and temporal information. However, numerous problems appear when discussions move from particles to extended bodies, such as the motion of wheels. At this junction, students experience profound difficulties if they continue to rely on the primitive causal schema to understand the new phenomena, because the motions of extended bodies can no longer be reconciled with the primitive causal schema. To understand the motion of a rolling wheel, for example, we must use property information such as the shape of the wheel, which defines the mass distribution of the subject. One of the most common mistakes that students make when they study motions of extended bodies is to ignore the critical property information and treat extended bodies as particles (Proffitt et al. 1990). In other words, these students fall back to a lower level of information processing and overextend the primitive causal schema to a domain that requires sophisticated analysis.

Even experts occasionally make the mistake of overextending the primitive causal schema to the domain of extended bodies. It is found that high school physics teachers and university physics professors exhibit confusions similar to those found in naïve students if they are forced to answer questions quickly or if they are not allowed to make explicit calculations. In an experiment, Proffitt and Gilden tested whether their subjects, 50 professors of physics, could correctly understand the motions of two wheels with identical mass and radius but different mass distributions (a solid wheel and a rimed wheel). They asked the subjects to predict whether the two wheels would arrive at the bottom at the same time when they were released simultaneously at the top of an inclined plane. Without making mathematical analyses, 80 % of the subjects incorrectly predicted that the wheels would roll down the inclined plane at the same rate, a performance similar to those of naive undergraduates. Apparently, when these experts of physics were prohibited from solving the problem analytically, they ignored the critical property information about mass distribution and fell back to the primitive causal schema that relies only on spatial and temporal cues. They failed to predict the result correctly, not because they appealed to erroneous theories, but because they appealed to the primitive causal schema that is intrinsically limited in dealing with simple mechanical systems. Interestingly, the subjects in the experiment continued to be perplexed even after observing how the wheels were actually rolling down the inclined plane, which suggests that the primitive causal schema is well entrenched, and adults, including experts, prefer to rely on this primitive schema whenever it is possible (Proffitt and Gilden 1989).

5 Conclusion

Findings from both developmental and cognitive psychology show that the evolution of causal knowledge has many important features different from what Piaget had described. First, as earlier as 6 months of age, infants are able to activate a primitive causal schema and use spatial and temporal cues to recognize causal relations in simple physical events such as mechanical collision. Second, as infants grow, they learn to use more complex information to identify causal relations, and they eventually acquire sophisticated schemas to represent causality. The evolution of causal knowledge consists in moving from processing simple, discrete and incoherent information to processing complex, holistic and coherent information. Third, the development of causal representations is not a smooth progression. The primitive causal schema never becomes irrelevant even after adults have acquired sophisticated schemas to understand complex causal relations. When adults are overloaded by processing complex information, they frequently fall back to a lower level of information process and overextend the primitive causal schema beyond the domain of mechanical causations. Misconceptions are bound to occur when people use the primitive causal schema to describe complex causal relations.

Misconceptions associated with the primitive causal schema have a unique cognitive mechanism. They occur because we activate an inappropriate schema that we had acquired at the very beginning of our cognitive development and use it to comprehend complex causal phenomena. Notice that we activate such an inappropriate schema neither because of insufficient cues, nor because of appropriate schemas being unavailable. Adults have already acquired mature and sophisticated causal schemas, and they fall back to the primitive schema only when their cognitive systems are overloaded by information (e.g. the input load is too high, the input contains a high level of noise or needs to be reclassified). Thus, falling back to a lower level of information processing is not a conscious decision—people return to the primitive causal schema with no awareness of an inappropriate one being activated. Strictly speaking, falling back to a lower, primitive level of cognitive processing is neither a strategy nor a heuristics, but an automatic, unconscious reaction. It is a habit of the mind. Because of this habit, we have a bias to the primitive causal schema, a preference to interpret all causal relations as direct and instant according to the framework of mechanical collisions.

The habit of falling back to a primitive information processing is well entrenched. Primitive information processing such as dealing with spatial and temporal cues requires minimal cognitive resources. Spatiotemporal information can be employed directly without any background or prerequisite information. At the same time, primitive information processing is usually effective—using merely spatial and temporal cues, infants can perform many task including identifying objects and recognizing mechanical causation. From a developmental perspective, primitive information processing in the form of spatial and temporal cues is fundamental. Spatiotemporal processing is the first cognitive capacity that infants acquire and it continues to function effectively and efficiently in adulthood. Thus, the mind has a natural reaction of falling back to a lower level of information processing when unfamiliar environment causes information overload.

Thus, it is not easy to correct misconceptions caused by an automatic, unconscious habit of the mind. Since the primitive causal schema is well entrenched and there is no a unidirectional disparity between successive schemas, it is difficult to prevent adults from falling back to the primitive causal schema and from using it improperly. We should expect that misconceptions associated with the primitive schema are widespread, across different ages, different educational levels, and different cultures. We should also expect that these misconceptions are persistent, because they are difficult to correct by instruction, and because they usually can't be overcome by a single attempt. To correct misconceptions associated with the habit of the mind represent a challenge.

Acknowledgments Preparation of this article was supported by a grant (13JZD004) from the Philosophy and Social Sciences Foundation of the Ministry of Education of P.R. China.

References

Aguiar, A., & Baillargeon, R. (1999). 2.5-month-old infants' reasoning about when objects. *Cognitive Psychology*, *39*, 116–157.

Baillargeon, R., Spelke, E., et al. (1985). Object permanence in 5-month-old infants. *Cognition*, *20*, 191–208.

Carey, S. (2009). *The origin of concepts*. Oxford: Oxford University Press.

Cohen, L., & Oakes, L. (1993). How infants perceive a simple causal event. *Developmental Psychology*, *29*, 421–433.

Cohen, L., & Amsel, G. (1998). Precursors to infants' perception of the causality of a simple event. *Infant Behavior and Development*, *21*, 713–732.

Cohen, L., & Cashon, C. (2001). Infant object segregation implies information integration. *Journal of Experimental Child Psychology*, *78*, 75–83.

Cohen, L., Chaput, H., et al. (2002). A constructivist model of infant cognition. *Cognitive Development*, *17*, 1323–1343.

Chi, M., Roscoe, R., et al. (2012). Misconceived causal explanations for emergent processes. *Cognitive Science*, *36*, 1–61.

Desrochers, S. (1999). Infants' processing of causal and noncausal events at 3.5 months of age. *The Journal of Genetic Psychology*, *160*.

Hespos, S., & Baillargeon, R. (2001). Reasoning about containment events in very young infants. *Cognition*, *78*, 207–245.

Kotovsky, L., & Baillargeon, R. (2000). Reasoning about collisions involving inert objects in 7.5-month-old infants. *Developmental Science*, *3*, 344–359.

Leslie, A. (1984). Spatiotemporal continuity and the perception of causality in infants. *Perception, 13*, 287–305.
Leslie, A., & Keeble, S. (1987). Do six-month-old infants perceive causality? *Cognition, 25*, 265–288.
Leslie, A., Xu, F., et al. (1998). Indexing and the object concept: Developing "what" and "where" systems. *Trends in Cognitive Sciences, 2*, 10–18.
Oakes, L., & Cohen, L. (1990). Infant perception of a causal event. *Cognitive Development, 5*, 193–207.
Piaget, J. (1954). *The construction of reality in the child*. New York: Basic.
Proffitt, D., & Gilden, D. (1989). Understanding natural dynamics. *Journal of Experimental Psychology: Human Perception and Performance, 15*, 384–393.
Proffitt, D., Kaiser, M., et al. (1990). Understanding wheel dynamics. *Cognitive Psychology, 22*, 342–373.
Rips, L. J. (2011). Causation from perception. *Perspectives on Psychological Science, 6*, 77–97.
Rakison, D., & Krogh, L. (2012). Does causal action facilitate causal perception in infants younger than 6 months of age? *Developmental Science, 15*, 43–53.
Spelke, E., & Kinzler, K. (2007). Core knowledge. *Developmental Science, 10*, 89–96.
Spelke, E., & Kestenbaum, R. (1986). Les origines du concept d'objet. *Psychologie Franaise, 31*, 67–72.
Spelke, E., Berinlinger, H., et al. (1992). Origins of knowledge. *Psychological Review, 99*, 603–632.
Spelke, E., Phillips, A., et al. (1995). Infants' knowledge of object motion and human action. In D. Sperber, D. Premack, & A. Premack (Eds.), *Causal cognition: A multidisciplinary debate* (pp. 44–78). Oxford: Clarendon Press.
Van de Walle, G., Rubenstein, J., et al. (1998). Infant sensitivity to show motions. *Cognitive Development, 13*, 387–419.
Xu, F. (1999). Object individuation and object identity in infancy: The role of spatiotemporal information, object property information, and language. *Acta Psychologica, 102*, 113–136.
Xu, F., & Carey, S. (1996). Infants' metaphysics: The case of numerical identity. *Cognitive Psychology, 30*, 111–153.

On the Nature and Composition of Abstract (Theoretical) Concepts: The X-Ception Theory and Methods for Its Assessment

Luigi Pastore, Sara Dellantonio, Claudio Mulatti and Remo Job

Abstract The 'standard picture of meaning' suggests that natural languages are composed of two different kinds of words: concrete words whose meaning rely on observable properties of external objects and abstract words which are essentially linguistic constructs. In this study, we challenge this picture and support a new view of the nature and composition of abstract concepts suggesting that they also rely to a greater or lesser degree on body-related information. Specifically, we support a version of this new view which we call "x-ception theory" maintaining that abstract concepts are based on internal information of a proprioceptive, interoceptive and affective kind. Secondly, we address a methodological issue concerning the so-called concreteness and imageability measures, two tools that are widely used in (mainly psycholinguistic) empirical research to assess the degree of concreteness of specific words. On the basis of this analysis we argue that—even though the classical concreteness and imageability measures were developed in relation to the standard picture of meaning—they can also be used in the new framework of x-ception theory. In particular, we suggest that the discrepancy between these two measures provides a clue as to whether a word relies on internal information. By contrast, we argue that a new measure for concreteness recently proposed in order to address some problems with the old measure is completely inappropriate for this aim.

L. Pastore
Università degli Studi di Bari, Bari, Italy
e-mail: luigi.pastore@uniba.it

S. Dellantonio (✉) · R. Job
Università degli Studi di Trento, Trento, Italy
e-mail: sara.dellantonio@unitn.it

R. Job
e-mail: remo.job@unitn.it

C. Mulatti
Università degli Studi di Padova, Padova, Italy
e-mail: claudio.mulatti@unipd.it

L. Magnani et al. (eds.), *Philosophy and Cognitive Science II*,
Studies in Applied Philosophy, Epistemology and Rational Ethics 20,
DOI 10.1007/978-3-319-18479-1_3

1 Introduction

The 'standard picture of meaning' suggests that natural languages are composed of two different kinds of words: concrete words "whose meaning are fixed by their relations with observable properties of the environment" and abstract words whose meanings are "fixed by a network of inferential or other relations to the meanings of other words, including those belonging to the observation vocabulary." (Cruse 2000, p. 52). According to this picture, these two word classes are structurally different from each other since abstract words are linguistic (i.e. purely definitional) constructs, while concrete words are based on perceptual information driven by the external world. Words that do not rely on perceptual information are organized in a structure of growing abstraction, in which terms based more directly on observational vocabulary are considered more concrete than those which rely on other linguistic constructs.

However, recent studies carried out mainly in the field of the so-called embodied cognition challenge this view and suggest that abstract words are not just linguistic constructs, and that at least some of them do rely on sensory information that is not driven by external perception, but rather concerns internal states of the body. As e.g. Barsalou states: "Recent embodiment theorists propose that knowledge acquired from introspection is central to the representation of abstract concepts." (Barsalou 2008, p. 620; for an overview see also e.g., Barsalou 1999; Barsalou et al. 2003; Barsalou and Wiemer-Hastings 2005). The view that abstract words rely (at least to some extent) on introspective information challenges this standard picture of meaning and reshuffles the cards of the classical architecture of concrete and abstract words.

In fact, some authors suggest that there is no strict opposition between concrete and abstract words, and that apparent differences result from the fact that words refer to concepts that are composed in different proportions of external sensory information, internal experience and linguistic information. As e.g. Vigliocco and collaborators maintain:

> The apparent dichotomy between concrete and abstract word meanings arises because of a statistical preponderance for sensory-motor information to underlie concrete word meanings and a preponderance for affective and linguistic information to underlie abstract word meanings. While sensory-motor information is statistically more preponderant for concrete word meanings, affective and linguistic information is statistically more important for abstract word meanings both for their acquisition and their subsequent representation in the adult system. (Vigliocco et al. 2009, p. 223; see also Kousta et al. 2011).

Not only has the standard picture of meaning been challenged by this view, but also the two measures psycholinguistics operationalized in order to distinguish between concrete and abstract words—the so-called 'concreteness' and 'imageability' constructs—seem to have become obsolete with respect to this debate, since they rely on the classical notion of concreteness as something based only on external perception.

In this paper we address both the theoretical and the methodological aspects of this issue. First of all, we discuss this new idea of 'abstractness' in comparison to the old one and we argue that word meanings reflect different degrees of involvement of different kinds of external and internal sensory information as well as of linguistic information, positioning words in different parts of a multidimensional space that allows clusters of words to be closer together on any of these dimensions. Specifically, we support a version of this new idea of 'abstractness' which we call 'x-ception theory' (Dellantonio et al. 2014) according to which abstract concepts rely to a greater or lesser extent on internal information of a proprioceptive, interoceptive and affective kind. This theory allows us—among other things—to give a more precise idea of what kind of abstract concepts might be based more heavily on internal information or might, on the contrary, rely on this information only to a minimal extent.

Secondly, we address a methodological issue concerning the so-called 'concreteness' and 'imageability' constructs, i.e. the main scales that have been defined to measure concreteness versus abstractness. Specifically, we will shed light on the definition of concreteness versus abstractness they are implicitly built on and illustrate what they really measure. On the basis of this analysis we argue that—even though the classical concreteness and imageability measures have been developed in relation to the standard picture of meaning—they can also be used in the new framework of x-ception theory. Specifically, we suggest that the discrepancy between the two offers a clue to assess whether a word relies on internal information. By contrast, we argue that a new measure for concreteness recently proposed in order to address some problems with the old measure, is completely inappropriate for this aim.

2 The Standard Picture of Abstractness

The classical view of abstraction proposed by both the philosophical and the psychological research defines abstract concepts (or words)[1] in opposition to concrete ones: a concept (or word) is considered as concrete if it denotes observable things in the external world, while it qualifies as abstract when it does not refer to something perceivable. This differentiation between concrete and abstract words/concepts is at the basis of the 'standard picture of meaning' according to which natural languages are composed of two structurally different classes of words, i.e. concrete words based on perceptual information "whose meaning are fixed by their relations with observable properties of the environment" and abstract words whose meanings are "fixed by a network of inferential or other relations to the meanings of other words, including those belonging to the observation vocabulary" which are therefore purely linguistic constructs (Cruse 2000, p. 52). Words that do not rely on perceptual information are organized in a structure of growing abstraction, in which those based more directly

[1]In our view, 'concepts' and 'words' are equivalent notions since concepts are considered to be the internal representations that support the semantic competence a person has with respect to the corresponding words. To know a word (i.e. to apply it correctly), one must have a corresponding concept that allows her/him to group together the class of objects denoted by the word. In this sense, concepts and words can be considered as equivalent at least in the sense that they must rely on corresponding inclusion/exclusion criteria (on this see also Dellantonio and Pastore 2006).

on observational vocabulary are considered more concrete than those which rely on other abstract linguistic constructs positioned higher in the structure.

As far as philosophical investigation is concerned, the standard picture of meaning originates in the field of philosophy of science and specifically from research on the relationship between the observable and unobservable entities in a theory and the nature of the terms describing those entities: the so-called *observational terms* denoting observable objects and so-called *theoretical terms* denoting unobservable entities (Carnap 1956; Achinstein 1965; Papineau 1996). There is no clear-cut distinction between theoretical and observational terms, but they are organized in a continuum from the perceptual (i.e. observational) periphery to the highest theoretical level. The more a term is distanced from the observational level the more reference fixing becomes problematic and dependent on the theory, i.e. on the place the term occupies in the corpus of the theory. Thus, theoretical terms depend heavily if not entirely on the definitional and mathematical apparatus of a theory and are related to the observational periphery only though the mediation of other terms. While the observational vocabulary can be considered quite stable intersubjectively, the theoretical vocabulary varies and changes in time together with (i.e. depending on) the theory that defines it.

These considerations regarding the observational and the theoretical vocabulary of a scientific theory apply in the same way to natural languages and to the differentiation between concrete and abstract words. In fact, the issue of theoretical terms addressed in the field of the philosophy of science is actually one and the same as the issue of abstract terms discussed in the field of psychology and, more recently, of philosophy of psychology. Concrete words are conceived as those that refer to perceptible, material entities that can be directly observed. As e.g. Jesse Prinz maintains, 'democracy' is an abstract concept because its referent cannot be directly experienced in perception; because it cannot be seen, heard, smelled, or tasted (Prinz 2002, p. 167). On the contrary, abstract words are analogous to theoretical terms in the sense that they do not have a referent that can be directly perceived, but they depend on a specific definition given by (some kind of) 'theory', which does not need to be a scientific theory, but can also consist in a commonsensical view. Think e.g. of words like 'democracy', 'truth' or 'belief' that are often presented as prototypical examples of abstract terms (see also e.g. Barsalou 1999; Connell and Lynott 2012): people have commonsensical views on the (definitions of) these things which determines the knowledge they have of the corresponding terms. Thus, what we identify as 'truth' or 'democracy' or 'beliefs' and the properties we associate with them depend (implicitly or explicitly) on the definition we acquired. As pointed out also by Alan Paivio—one of the most important authors in psychology who tried to account for the difference between abstract and concrete terms from the point of view of the psychological mechanisms that are supposed to underlie their processing—the philosophical discussion on theoretical terms is closely related to the psychological discussion on abstract words: "My perspective on the issue is pragmatic and psychological. The basic assumption is that the observational-theoretical distinction becomes psychologically real when interpreted in terms of the correlated difference between concrete and abstract terms." (Paivio 1986, p. 11).

When Paivio says that the distinction between abstract and concrete terms is "psychologically real" he is not just speaking metaphorically to imply that this is psychologically relevant. In fact, he is rather referring to a specific hypothesis he puts forward on the way abstract terms are mentally represented which is called *Dual Coding Theory*. At its core, Paivio's Dual Coding Theory suggests that the cognitive system is composed of two different symbolic subsystems connected to each other: a verbal one, specialized for the processing of language, and a nonverbal one, also called the imagery system, whose function is mainly to perceptually analyze the external world and generate mental images of it. The word 'imagery' describes both this system and the capacity this system enables to dynamically form and recall mental images. Mental images are not meant as visual pictures only, but basically describe traces stored in memory of all kind of sensations—acoustic, olfactory, haptic and gustatory. In this sense, an 'acoustic image' would be the nonverbal representation of a sound, an olfactory image would be the nonverbal representation of a smell, etc. As Paivio specifies: "Our minds 'contain' memory isomorphs of how entities and events look, sound, feel." (Paivio 2007, p. 25).

According to the Dual Coding Theory, the main difference between concrete and abstract words is that, while concrete words are represented in both the verbal and the nonverbal systems, abstract words are represented in the verbal system only: they depend largely on linguistic information and have only weak referential relationships.

> Both classes of words (abstract and concrete) have interconnections with the representations of other words in the verbal system. The specific nature and structure of the verbal-associative networks for concrete and abstract words presumably differ in systematic ways that reflect differences in the contexts in which they have been acquired and used, but in general it can be said that concrete and abstract words are semantically differentiated by the degree of availability of referential interconnections. Concrete words have both referential and verbal-associative meaning, whereas abstract words depend relatively more on verbal-associative interconnections for their meaning. (Paivio 1986, p. 123).

For this reason "comprehension is more dependent on imagery in the case of concrete than abstract sentences." (Paivio 1986, p. 219).

Paivio's hypothesis is part of a more general line of research on the so-called mental imagery which investigates the processes and the mechanisms through which we can mentally recall any kind of sensation and that support capacities that are usually described as 'visualizing', 'seeing in the mind's eye', 'hearing in the head' or 'imagining the feel of something'. Paivio's proposal was—as in general most theories supporting the existence of mental imagery developed in the same years (for an overview see e.g. Nigel 2013)—a reaction against the classical computational view according to which linguistic representations need to be conceived uniquely as abstract, amodal and language-like symbolic structures unrelated with the physical and functional features of the referents and not bound to the perceptual system. Computational views explain conceptual thought and linguistic capacities on the basis of internal relations between symbols, without adequately accounting for the problem of how these are connected to the external world. Imagery is introduced to explain how such amodal symbols are anchored to nonlinguistic perception so that people can understand in a referentially salient sense the meaning of words. In this sense, the main point of

Paivio's Dual Code might be seen in the problem of understanding how reference works: how people identify what things in the word correspond to a certain word. The implicit idea of a Dual Code is that—even though language understanding is partially due to verbal-associative interconnections and among verbal representations—these verbal-associative interconnections are not enough to account for the human capacity of using words to denote things. Thus, the problem Paivio faces on the basis of his theory is to account for both aspects of language understanding, the capacity to define words through other words and the capacity to connect words with things in the external word.

In this respect, Paivio's theory complies with the standard picture of meaning in suggesting that language is characterized by a dichotomy between terms that do have an observable reference in the external world which can be perceived through the senses and terms that are based on definitions. Thus, both the classical philosophical and psychological views on language and specifically on semantics rely on the idea that the ingredients needed to fix the meaning of terms and to understand them are sensory information and linguistic knowledge only. No other ingredient seemed to be needed. Abstractness and concreteness were defined on the basis of these two kinds of information only and their degree was supposed to depend on their reciprocal weights. This is the 'transdisciplinary mainstream' in opposition to which a new perspective has recently begun to be developed. This new perspective—which we are going to present in the next section—challenges this two-dimensional picture of language and frames a three-dimensional model, which does not only include external sensory information, and verbal-associative information, but also internal bodily experience.

3 Abstractness in a New Perspective: The X-ception Theory

The idea that the referential component of meaning can be explained solely in terms of the relationship between words and something observable in the external world is a classical principle not only in psychological but also in philosophical research. As for the latter, there is an influential research tradition arguing that—in order to ensure the intersubjectivity of meaning—reference must consist of something that everybody can observe. If the reference were 'private' and perceivable by a person only, we would never be sure that different people use the same words to indicate the same things (Wittgenstein 1953, 1967, §256ff; Kripke 1982; for an overview on the so-called 'private language argument' see e.g. Cook 1965; Schroeder 1998; Knorrp 2003; Candlish and Wrisley 2012). For this reason, according to this tradition the reference of words cannot be identified using internal sensory experience since this would be accessible only by the first person and stability of meaning couldn't be granted: i.e. it could neither be granted that words have univocal meanings independently of the specific person who uses them, nor that everybody uses them to denote one and the same thing.

However, the first attempts to assess Paivio's theory indicated that taking the position that reference can be understood uniquely as a relationship between a word and some observable object or property in the external world might not hold. The element in Paivio's theory that immediately presented difficulties was the emotions (specifically, emotion words). In fact, since emotions do not have any observable reference in the external world, they should not be represented in the imagery system, and the meaning of emotion words should be understood uniquely on the basis of verbal-associative interconnections. However, in Paivio's study of 1968 he had already pointed out that the situation seems to be more complex than this.

According to Paivio's theory, words that easily evoke sensory information have a direct representation in the imagery system, while words that evoke sensory information only with difficulty are represented solely in the verbal system. Since Paivio hypothesized that only concrete words have a direct representation in the imagery system, he predicted that only this word class should easily arouse sensory information. To confirm this hypothesis, Paivio developed an imageability scale that measures the image-evoking capacity of words meant as their capacity to arouse sensory (i.e. perceptual) experience stored in memory (Paivio 1965; Paivio et al. 1968) and then confronted the ratings obtained using this scale with the concreteness ratings previously collected for the same words by Spreen and Schulz (1966). If Paivio's model made the right prediction, imageability should always reflect concreteness and the two ratings should always be strongly correlated. Paivio et al. (1968) found indeed a high correlation (0.83) which was widely confirmed by later studies. In a recent analysis on the ratings of 4260 words included in the MRC database (Coltheart 1981; Wilson 1988) we confirmed a positive correlation between imageability and concreteness of 0.835 (Dellantonio et al. 2014; in this paper we also report other literature confirming this result). However, in spite of this strong general correlation, the prediction is not entirely confirmed since there are a number of words mainly referring to emotional states for which imageability and concreteness are uncorrelated. For those problematic items, imageability ratings are significantly higher than concreteness ratings. As Paivio observed, these items exhibit interesting similarities:

> Most of these are words with strong emotional and evaluative connotations. The largest group consists of terms referring to affective reactions or affective attitudes: AFFECTION, AGONY, AMAZEMENT, AMOUR, ANGER, ANXIETY, DEVOTION, FUN, GAIETY, GRATITUDE, GRIEF, HAPPINESS, HATRED, HOPE, HOSTILITY, HUMOR, INSOLENCE, JEALOUSY, JOVIALITY, JOY, KINDNESS, LOVE, LOYALTY, MISERY, MOOD, PANIC, PASSION, PLEASURE, PRIDE, SADNESS, SHAME. Others in this category are labels for attitudes and emotional situations, or are generally evaluative in meaning: BLESSING, BRAVERY, CHAOS, CHARM, CHRISTMAS, COURTSHIP, DEATH, GLORY, OBEDIENCE, OBSESSION, SAFETY, TRAGEDY and VANITY. (Paivio et al. 1968, p. 7).[2]

Analogous inconsistencies were noted in later studies (see e.g. Altarriba et al. 1999; Altarriba and Bauer 2004, p. 397). We were also able to confirm them in a previous study (Dellantonio et al. 2014) conducted on the concreteness and imageability

[2]The few anomalies in the opposite direction are quite easily explained: they involve words like 'antitoxin', 'armadillo', 'encephalon', 'dell' which denote concrete objects that are unusual and therefore most of the raters are not familiar with their appearance.

ratings collected in the MRC database which is one of the most important available sources for imageability and concreteness ratings (Coltheart 1981; Wilson 1988). In this work we selected 36 emotion and moods words taken from a number of studies on basic emotions and moods (Tomkins 1962, 1963; Ekman et al. 1969; Plutchik 1980; Ekman 1994, 1999; Reizenzein 2009; Kassam et al. 2013; Prinz 2004; Damasio 1999) and we compared them with ten randomly chosen control groups. The results showed that imageability ratings were significantly higher than concreteness ratings for emotion/mood words compared to the words in control groups.

The problem that Paivio already identified with respect to these recalcitrant words is that—even though they surely do not denote anything observable and therefore cannot be qualified as concrete—they still rely on some kind of sensory experience: "These words appear to have the common property of having been associated with sensory experience (usually affective in nature) but not specific things or classes of things". (Paivio et al. 1968, p. 7). This point raised by Paivio is as crucial as it is controversial since it suggests that word meanings might not be based on external perception and linguistic information only, but they might also *be grounded in another kind of sensory experience, i.e. in affective experience.*

If this is correct, then the reason why emotion words have a high imageability (i.e. according to Paivio's definition of imageability: easily arouse sensory experience stored in memory) is that they are represented in the non-verbal, imagery system like concrete words and that they have therefore a referential component, consisting in the affective experience that characterizes their occurrence. Thus, it is plausible to assume that the inconsistency between concreteness and imageability ratings in the case of emotion words is not a problem of the theory, but provides a clue to an unexpected situation. Specifically, it may indicate that the understanding of non-concrete words (i.e. words with low concreteness ratings) like emotion words also relies on sensory experience stored in memory, which is however not of an external, but of an internal kind. If this is the case, imageability should be interpreted as *a measure of the ease/difficulty with which a word evokes both external and internal sensory experience.*

As for concrete words, imageability correlates with concreteness because concreteness measures the link between a word and some external sensory information while imageability assesses the easiness/difficulty with which a word evokes this external sensory experience stored in memory. In the case of non-concrete words, imageability does not necessarily correlate with concreteness, because it is possible that—as in the case of emotions—even though a word is not linked to external sensory information, it relies on internal sensory experience. *If so, then a discrepancy between imageability and concreteness ratings like that shown by emotion words can be considered an important clue that a word easily evokes internal sensory experience* (for a more detailed discussion of this hypothesis see Dellantonio et al. 2014).

This conclusion is consistent with many recent theories, mainly those related to the tradition of embodied cognition which maintain that "knowledge acquired from introspection is central to the representation of abstract concepts." (Barsalou 2008, p. 620; see also e.g., Barsalou 1999; Barsalou et al. 2003; Barsalou and Wiemer-

Hastings 2005; Lakoff 1987; Johnson 1987; Lakoff and Johnson 1980; Gibbs and Jr 1994; Lakoff and Turner 1989). As Barsalou specifies elsewhere, introspective states "include events perceived inside the mind and body that typically lack counterparts in the external world, such as emotions, affects, appetitive states, cognitive operations, and beliefs." (Barsalou et al. 2003, p. 44 nota). As also e.g. Prinz points out, these states "stem [· · ·] from within the body (as with proprioception, interoception, hunger, and thirst)." (Prinz 2002, p. 116).

One of the most recent attempts made by the psychological (i.e. psycholinguistical) research to theoretically systematize this conception of abstract terms and their mental representation is that proposed by authors like Kousta and Vigliocco (see e.g. Vigliocco et al. 2009; Kousta et al. 2011). According to them, both concrete and abstract words are made of two different types of information: 'experiential information' and 'linguistic information'. The 'linguistic information' is of verbal-associative kind and it corresponds to the one the standard picture relies on. As for 'experiential information', they think that this is of a twofold kind, or more specifically that it is drawn from two different sources depending on whether we are considering concrete or abstract concepts. The experiential information concrete concepts consist in is described as 'sensory-motor information': this notion is meant to indicate perceptual information driven by sensory-motor interactions with the outside world. The experiential information abstract concepts rely on is identified as 'affective information', i.e. information driven by the experience everyone has of their emotions and emotional states. (Vigliocco et al. 2009). As e.g. Vigliocco and collaborators maintain:

> The apparent dichotomy between concrete and abstract word meanings arises because of a statistical preponderance for sensory-motor information to underlie concrete word meanings and a preponderance for *affective* and linguistic information to underlie abstract word meanings. While sensory-motor information is statistically more preponderant for concrete word meanings, *affective* and linguistic information is statistically more important for abstract word meanings, both for their acquisition and their subsequent representation in the adult system. (Vigliocco et al. 2009, p. 223—italics added)

And further, in the version of this view proposed by Kousta and collaborators:

> [...] we propose that both concrete and abstract concepts bind different types of information: experiential information (sensory, motor, and *affective*) and also linguistic information. However, concrete and abstract semantic representations differ in terms of whether sensory, motor, or *affective* information have the greatest weight, with sensory-motor information being more preponderant for concrete concepts and *affective* information playing a greater role for abstract concepts. Thus, a central and novel element of this proposal is the idea that experiential information contributes to the representation of both concrete and abstract words. However, whereas sensory-motor information is statistically more important for the representation of concrete words, *emotional content*, a largely neglected type of experiential information in the literature on semantic representation/processing, contributes to word representation and processing, particularly for abstract concepts. (Kousta et al. 2011, p. 14—*italics added*)

These quotes show that Vigliocco, Kousta and collaborators use the word 'affective' as a synonym for 'emotional',[3] thus considering emotional content as the only kind of internal experience abstract concepts rely on. However, *affective* experience meant in the sense of emotional experiences is only one of the various kinds of internal experience subjects perceive 'introspectively' (on introspection as a kind of perception see Goldman 1993). As implied by Barsalou's and Prinz's quotes reported above, there are a variety of other types of internal information we can introspectively access.

Other types of internal, introspective states very similar to emotions which also rely on sensory experience are those conveyed by *proprioception* (which provides information on body position, body movements and the muscular system, see e.g., Berthoz 2000) and *interoception*[4] (which provides information on the internal condition of the body: monitoring states like heartbeat, respiration, pain, hunger, thirst, the need for digestion, elimination, etc., see Craig 2003, 2009, 2010 for a more specific discussion of proprioception and interoception as well as of their relationship to emotional experience see Dellantonio, Pastore, forthcoming, Chap. 1 §5, §6).[5] Since there is no reason to think that affective experience is special with respect to proprioception and interoception and that affective experience is therefore *the only* relevant,

[3]In the philosophical and in the classical psychological tradition, terms like "affect", "affective" or "affection" assume a different meaning. In fact, the word "affection" is derived from Latin terms like *affectus* and *afficere* used to translate the Greek term *pathos*" which indicates the experience of any kind of event or modification caused by the interaction with an entity other than myself. The verb "to affect" in English preserves in part this original meaning. The word is used in this way e.g. by Aristotle in his doctrine of categories (Ackrill 1963). However, in his *On the Soul* Aristotle also used the word "affection" in another more restricted and specific sense to indicate only the passive modifications of the psyche that occur without the active and voluntary participation of the subject. In this sense "affections" are identified with and described as *passions*. In the modern era the word "affection" was used in this restricted sense by Descartes (Brown 2006) and Spinoza (Lebuffe 2010). This interpretation of "affection" in the sense of passion is also that inherited by psychological research. Here "affection" indicates not only the passive modifications of the psyche, but also the effects of this alteration (Dixon 2003). Thus, the word implicitly recalls the idea of an effect caused by some modifications. In a non-metaphysical context these modifications must concern primarily the body, thus we should use "affection" to indicate any state which is the effect of some bodily modification. In this sense, interoception and proprioception are univocally particular kinds of affective states while emotions belong to the class of affections and are affective states only insofar as they are caused by bodily modifications. However, this sense of the word has been lost in the contemporary psychological usage of the term in which affective and emotional have become synonyms.

[4]To be more precise, we should distinguish between interoception and nociception, which is the perception of any kind of pain; however, as e.g. Craig (2003, 2009) and Damasio (1999, 2010) maintain, nociception can be considered a form of interoception.

[5]Interoceptive and proprioceptive information might even be essential for the perception of emotional feelings since we experience our emotions also by means or in virtue of specific corresponding bodily changes: for example we experience fear also by virtue/means of bodily changes such as increased heart and respiration rates, muscular tension etc.; we experience shame also by virtue/means of increased blood flow in the face resulting in facial blushing, gastric and visceral contractions, etc.). On the relationship between emotional information and interoceptive/proprioceptive information see Dellantonio, Pastore, forthcoming, Chap. 4).

internal source of sensory information for the understanding of word meaning, it is plausible to hypothesize that the view proposed by Kousta, Vigliocco and collaborators might be extended at least to *interoceptive and proprioceptive experience. This would imply that word meanings might rely on all kinds of internal information people consciously perceive on the internal states of their body including (at least) affection, interoception and proprioception. We called this x-ception theory* (Dellantonio et al. 2014).

The idea that the representations of words denoting proprioceptive or interoceptive states relies on the corresponding internal information was the object of a previous study we carried out, in which we analyzed the imageability and concreteness ratings of a set of words included in the MRC database which denote proprioceptive and interoceptive states (Dellantonio et al. 2014). Our statistical analysis confirmed that this word class exhibits properties analogous to emotion words: i.e. it proved that the difference between imageability and concreteness ratings for interoceptive and proprioceptive words is significantly higher than that for the control groups (randomly chosen from the database). Also in this case, the inconsistency between imageability and concreteness ratings might indicate that—even though these words are not linked to external sensory information—they still arouse internal sensory experience.

As a further proof of the hypothesis that the difference between imageability and concreteness can be interpreted as a measure of whether a word representation relies on internal information, another data set analyzed in the previously mentioned study (Dellantonio et al. 2014) is worth reporting. Here we selected a number of theoretical terms relying on the definition of theoretical given by the classical philosophical debate (see §2); specifically, we chose abstract terms belonging to the technical jargon of a discipline whose meaning is highly dependent on the linguistic definition given them by the framework of that theory like e.g. 'axiom' (mathematics); 'causality' (physics); 'conjugation' (linguistics); 'legislation' (politics and law); 'deduction' (logic); or 'theory' (science in general). Our hypothesis suggests that for this particular class of abstract words, no sensible discrepancy between imageability and concreteness should be observed, since this word class should not rely (at least not to a large extent) on internal sensory information. In fact, as in the case of concrete words, their imageability ratings should depend on their concreteness ratings only and the two measures should correlate. In Dellantonio et al. (2014), we found that the differences between imageability and concreteness for the theoretical/technical words are either smaller than, or comparable to, those of the control groups. In addition, we carried out another analysis specifically for this paper confirming that they are also correlated ($r = 0.743$, $p < 0.001$).

Since the method we developed to assess the internal grounding of words based on the discrepancy between imageability and concreteness ratings gave encouraging results, we used it to provide some new data for this study. Specifically, we tested Barsalou's hypothesis (Barsalou 1999; Barsalou et al. 2003) that our capacity to understand words denoting what can be termed doxastic mental states such as e.g. 'to know', 'to believe', 'to be certain', 'truth', 'false' etc. and words denoting states like 'hope', 'desire', 'wish', 'remember', etc. which we might more properly call attitudes is due, at least in part, to the availability of internal, introspectively accessible information for these states. To do so, we analyzed a new group of words included in the MRC database denoting specifically doxastic and epistemic states (e.g. ‘true’, ‘truth’, ‘right’, ‘wrong’, ‘certain’ ‘uncertain’, ‘falsehood’, ‘unknown’, etc.). Even though the number of words we were able to include in the analysis is relatively low (16), the analysis showed that the mean difference between imageability and concreteness of doxastic states words (mean difference = 60) is statistically identical to the mean difference between imageability and concreteness of x-ceptive words (mean difference = 72), $t < 1$. More specifically, doxastic words behave like proprioceptive and interoceptive words.[6]

This result is particularly interesting since it shows that there are also other classes of non-concrete words beyond proprioceptive, interoceptive and emotional terms that rely on internal information and this offers some support for the thesis that many if not all abstract (i.e. non-concrete) concepts include some internal information. As further proof of this and to see what kind of concepts are more largely based on internal information we used this method of identifying the discrepancy between imageability and concreteness in the opposite direction. Instead of starting from specific terminological classes selected a priori on the basis of some criteria, we extracted all the words that have the highest discrepancy between imageability and concreteness from the MRC database: for each word in the MRC database provided with imageability and concreteness ratings we computed the difference between imageabiilty and concreteness, we then computed the mean and standard deviation of this set and selected those words for which the difference between imageability and concreteness was 2 standard deviations above the mean of the database. The words selected on the basis of this criterion are interestingly varied, confirming that internal information is crucial not only for the understanding of proprioceptive, interoceptive and emotional terms, but that many word types largely rely on introspective information.

In fact, among the term matching this criterion there are not only the words classes we expected to find on the basis of our results, i.e. words denoting:

- *emotions/moods* like ‘joy’, ‘jealousy’, ‘happiness’, ‘love’, ‘unhappiness’, ‘fun’, ‘optimism’, ‘terror’ etc.;
- *interoceptive* and *propriceptive states* such as ‘relaxation’, ‘warmth’, ‘excitement’, ‘thrill’ etc.;
- conditions that can be considered *midway between interoceptive and emotive states* such as ‘pleasure’, ‘anxiety’, ‘tranquil’, ‘excitement’, ‘unpleasantness’, ‘uneasiness’ etc.;
- or *states closely related to emotions* such as ‘grief’, ‘hostility’, ‘bravery’, ‘romance’, ‘intimate’, ‘danger’, ‘humor’, ‘seduction’, ‘beauty’.

This group of words also includes other kinds of terms that are usually more univocally considered as abstract like words denoting:

---

[6]In this regard, it should be specified that the mean difference between imageability and concreteness of proprioceptive and interoceptive terms is significantly smaller than the mean difference between the imageability and concreteness of emotion words (mean difference = 119), $t(47) = 2.8$, $p < 0.01$. We do not have a conclusive explanation for this result, however see in Dellantonio et al. (2014) for some possible hypotheses.

- *supernatural/religious phenomena* such as 'devil', 'mystery', 'magic', 'ghost', 'eternal', 'devotion', 'goddess', 'demon', 'sin', 'hell', 'angel', 'paradise';
- *significant events or times in life* such as 'maternity', 'marry', 'graduation', 'marriage', 'holiday', 'adolescence', 'maturity', 'childhood', 'summer';
- *interpersonal behaviors related to values* such as 'kindness', 'insolence', 'hostility', 'obedience', 'adultery', 'gratitude', 'greed', 'luxury', 'vanity', 'independence', 'pride', 'failure';
- *emotionally connoted social relationships* such as 'marriage', 'freedom', 'friendship', 'poverty';
- *mental states* such as 'obsession', 'delirium', 'hope', 'reflection'.

If our interpretation of the imageability measure and of its relationship with the concreteness measure is correct, then these words are strongly related to internal sensory information of some kind. We neither have a specific explanation for why these words seem to be more linked than other words to internal information, nor do we completely trust the accuracy of the measure we use (about this see the next section). However, this result univocally indicates that many words commonly considered as abstract seem—according to the imageability ratings people assign to them—not to be merely linguistic constructs, but to rely on some kind of internal information. What kind of internal information they rely on must be the object of further investigation.

In spite of this open question, the analysis we carried out as well as the data we reported allow us now to draw some conclusions regarding two aspects we considered previously. On the one hand, these findings challenge the classical view of abstract words according to which they must be conceived as theoretical terms, i.e. as purely linguistic constructs. They suggest instead a differentiation of degree between abstract words which are closer to theoretical terms as they are classically defined (those abstract terms whose representations consist almost uniquely of verbal associative information) and abstract words of a different kind, whose representations rely for a large part also on internal sensory experience.

On the other hand, on the basis of our observations there is no reason to maintain that affective experience is the only kind of internal information that plays a role with respect to our understanding of abstract terms, as implicitly suggested by Vigliocco, Kousta and collaborators. On the contrary, our findings lead us to think that all kinds of internal experience introspectively available to the subject might be relevant for understanding (some) non-concrete words. At the very least, proprioceptive and introspective information are surely part of the internal information people rely on to understand some word classes. For this reason we call our view x-ception theory.

By suggesting that a more accurate classification of the internal information people have at their disposal is essential to shed light on the debate on the semantic differences among words and words representations, we propose a view according to which word meanings reflect different degrees of involvement of different kinds of external and internal sensory information as well as linguistic information, positioning words in different parts of a multidimensional space that allows clusters of words to be closer together on any of these three dimensions. In consideration of this position, the very notion of reference needs to be revisited, taking into account the possibility that at least some words might have some form of internal reference: i.e. that they might denote in a more or less direct way internal states, which are introspectively perceived by the first person and that the sensory experience of these states is part of the information our semantic representations consists in.

This view might have relevant consequences also with respect to our idea of concreteness, because it challenges the classical dichotomy between concrete words that are grounded in external information and abstract words that are understood primarily on the basis of linguistic information and it suggests that some abstract words are also grounded in some sort of sensory information, even though of an internal kind. In fact, by virtue of this 'grounding', these abstract words might be in some respects more similar to concrete words that to theoretical words. In particular, it is possible that the so-called concreteness effect (claimed to be responsible for the fact that concrete terms are processed more quickly than abstract terms[7] and are learned earlier and more easily than abstract terms) does not only apply to words that rely on external sensory information. In fact, this might also take place in the case of words linked to internal sensory information, even though to a lesser extent since internal experience is less specific and less univocal than perceptual information. This is an empirical hypothesis which needs to be further investigated.

4 Methodological Issues: Is There a Way to Assess Internal Grounding?

The idea that introspective information is central to the representation of abstract concepts reshuffles the cards of the classical architecture of concrete and abstract words, challenging the standard picture of meaning according to which word representations are either grounded in external sensory information or in linguistic constructs, and suggests that words may also rely on internal sensory information.

Because of this radical perspective change, the measures psycholinguistics operationalized in order to distinguish between concrete and abstract words—the so-called 'concreteness' and 'imageability' constructs—could be obsolete since they were developed relying on the standard picture of meaning. However, the analysis we carried out in the previous part shows that this is not the case, at least not entirely. In fact, in §3 we suggested that the discrepancy (i.e. difference) between imageability and concreteness ratings offers an index of whether a word representation relies on internal information. However, this does not mean that these scales are perfectly adequate to differentiate between different kinds of non-concrete words and to establish which are mainly linguistic constructs and which are internally grounded. In this section we discuss some issues related to these two measures and their potential

[7]This means that the reaction times in tasks like lexical decision, word naming and recall are shorter for concrete than for abstract terms. For a recent review of the literature on this effect see e.g. Connell and Lynott (2012).

with respect to the new view. More specifically, we explain: (i) why the old constructs defined by Spreen and Schulz (1966) (concreteness) and Paivio et al. (1968) (imageability) that we used in our studies (this one as well as Dellantonio et al. 2014) are indeed appropriate for indicating whether words are internally grounded, even though they are not completely reliable and (ii) why on the contrary the new construct of concreteness defined by Brysbaert et al. (2014) is not at all reliable for our purposes.

First of all, as we have discussed extensively elsewhere, it is only a 'methodological accident' that the imageability construct defined by Paivio is also a measure of the internal grounding of words. This accident is due to the fact that the instructions given to the participants on how to assign imageability ratings were misleading (Dellantonio et al. 2014):

> Nouns differ in their capacity to arouse mental images of things or events. Some words arouse a sensory experience, such as a mental picture or sound, very quickly and easily, whereas others may do so only with difficulty (i.e., after a long delay) or not at all. The purpose of this experiment is to rate a list of words as to the ease or difficulty with which they arouse mental images. Any word which, in your estimation, arouses a mental image (i.e., a mental picture, or sound, or other sensory experience) very quickly and easily should be given a high imagery rating: any word that arouses a mental image with difficulty or not at all should be given a low imagery rating. Think of the words 'apple' or 'fact'. Apple would probably arouse an image relatively easily and would be rated as high imagery; fact would probably do so with difficulty and would be rated as low imagery. (Paivio et al. 1968, p. 4)

As we mentioned in the previous section, Paivio thought that imageability should closely resemble concreteness in all cases because in his view only concrete words had references that could be represented non-linguistically in the imagery system and were therefore easy to imagine (i.e. have high imageability ratings). For this reason, Paivio designed the instructions for the collection of imageability ratings primarily in terms of the ease or difficulty with which words arouse mental images of *external things or events* and the examples he uses are 'apple' and 'fact'.

However, in Paivio's view everything that can be perceived through the senses including e.g. smells, tastes, voices etc. is concrete. Thus, his notion of 'mental image' describes traces stored in memory of all types of external sensations (not only visual ones, but also auditory, olfactory, gustatory and haptic sensations should be taken into account). It is for this reason that in the instructions he specifies that all kind of sensory experience ("a sensory experience"; "other sensory experience") should be considered when making an imageability judgment. However, the request to estimate imageability depending on whether/how much a word arouses sensory experience without further specifications might have lead participants to assign their ratings on the basis of the ease/difficulty with which words aroused any kind of sensory experience stored in memory, including internal, body-related sensations. Thus, the fact that imageability appears to indirectly measure also the internal grounding of words (when compared with concreteness) is the side-effect of ambiguous instructions making the instructions instrumental in collecting biased ratings.

Even though the instructions of imageability suggested evaluating the ease/difficulty with which a word evoked any kind of sensory information including possibly internal experience, they are certainly biased towards an idea of imageability that is primarily visual and related to the ease/difficulty with which people can form a mental picture of the referent of a word. In fact, the term 'image' recalls quite strongly the idea of a visual picture. Thus, this instruction is certainly biased towards the sense of vision. This bias could represent a problem when we compare imageability with concreteness ratings. However, in the following part we will show that it does not. Specifically, we will argue that the instructions for concreteness also favor the sense of vision and that these two biases 'balance each other out'.

(i) The original instructions for concreteness ratings developed by Spreen and Schulz, (1966, p. 460) were as follows:

> Nouns may refer to persons, places and things that can be seen, heard, felt, smelled or tasted or to more abstract concepts that cannot be experienced by our senses. The purpose of this experiment is to rate a list of words with respect to 'concreteness' in term of sense-experience. Any word that refers to objects, material or persons should receive a high concreteness rating; any word that refers to an abstract concept that cannot be experienced by the senses should receive a low concreteness rating. Think of the words 'chair' and 'independence'. 'Chair' can be experienced by our senses and therefore should be rated as high concrete; 'independence' cannot be experienced by the senses as such and therefore should be rated as low concrete (abstract).

According to the concreteness instructions something is concrete if it can be perceived through (at least one of) the senses. However, as it is has been already pointed out (Connell and Lynott 2012, p. 461), the examples mentioned in the second part of the definition ("objects, material or persons" as well as "chair" versus "independence") might have been misleading. In particular, they might have biased people to rely for their ratings (also) on a different idea of concreteness which resembles more closely the everyday understanding of the word 'concrete' and its dictionary definition, according to which 'concrete' means material or physical and an object is concrete only if it has a material composition. Since material objects are perceived mainly or primarily through vision and possibly through touch, people's ratings probably favored these senses over the others. However, since external material things are the ones people can more easily form a mental picture of, it can be predicted that for these material things there is an overlap between concreteness and imageability, independently of the sensory channel through which such material things are perceived.

We tested the hypotheses derived from the perusal of the instructions by performing two analyses. In the first analysis we checked whether there was a bias toward the sense of vision in Spreen and Schulz (1966) participants' ratings. We took into consideration the words included in the MRC database that have a concreteness rating (N = 4260). We ordered them in descending order from those with the highest concreteness ratings to those with lower rating and determined if they were related to the sense of vision or to some other senses. All the words up to the 705th item, i.e. all words whose concreteness ratings exhibit a standard deviation higher than 1,17, referred to material objects that could be perceived by vision.

In the second analysis we checked whether the imageability ratings for words referring to senses other than vision did indeed correlate with the concreteness ratings for the same words. To do so we took into consideration the concreteness ratings of a list of words included in the MRC database denoting sounds, tastes and smells like 'sound', 'voice', 'rumor', 'salty', 'spicy', 'bitter', 'scent', 'odor', 'smell' etc. (words connected to touch like smooth, rough, soft, tender were not considered since they have more ambiguous meanings) and compared them with the mean concreteness ratings of other items in the corpus (purged of words occurring multiple times.) Things that can be perceived through senses other than sight (N = 20) received mean concreteness and imageability ratings (432 and 460 respectively) comparable to those of the corpus as a whole (N = 4239; 439 and 456 respectively). Moreover the correlation between imageability and concreteness for those words is significant ($r = 0.824$, $p < 0.01$). This comparison confirms that words denoting sounds, tastes and smells received only medium concreteness ratings and were therefore considered as less concrete than words denoting material objects that can be perceived through vision. However, the high correlation between concreteness and imageability ratings also points out that this class doesn't exhibit anomalies analogous to those observed with regard to emotion words, and it is therefore perfectly consistent with Paivio's view.

The results from our analyses can be summarized as follow. Despite what some studies maintain (e.g. Vigliocco et al. 2009; Connell and Lynott 2012), the ease/difficulty with which a word evokes a mental picture of a visual kind is not the only relevant aspect measured by imageability. Imageability also tracks the ease/difficulty with which a word evokes mental images aroused by senses other than vision. In fact, our analysis of words denoting sounds, tastes, and smells (we couldn't check touch) shows that—even though these imageability ratings tend to be lower than those for words denoting material objects—they are still highly correlated with concreteness ratings. Our data indicates that words denoting sounds, tastes, and smells are perceived as a bit less concrete and as a bit less imageable than words denoting material objects that can be seen. There is no inconsistency between the perceived concreteness and the perceived imageability in such cases (the correlation is significant); thus, the bias toward vision is not relevant for our conclusions concerning the viability of using the difference between imageability and concreteness as a measure of the internal grounding of a word.

Even though this bias is not relevant from our point of view, other authors have considered it extremely relevant and argue that many problems found in psycholinguistic studies in the past can be traced back to this bias. They therefore propose that a new concreteness measure should be defined to overcome this obstacle. One of the most influential studies with this aim is Connell and Lynott (2012). The authors' primary interest concerns the concreteness effect (mentioned at the end of §3), i.e. the behavioral advantages exhibited by words referring to concrete objects that are processed more quickly and accurately than abstract words. They point out that "despite their reputation as a textbook effect, concreteness effects do not always reliably emerge in semantic processing" (p. 453). In their view these problems are not due to the effect itself (which they consider indeed to be absolutely reliable), but rather to the means that are typically used to select the items for the experiments, i.e. the concreteness and the imageability measures. They think that both the concreteness and imageability ratings (which are often used interchangeably to assess the degree of concreteness versus abstractness of words) do not adequately capture the degree to which terms rely on perceptual experience because both the instructions for concreteness and imageability are affected by the biases we mentioned above. Specifically, while imageability is affected by a visual bias, concreteness is evaluated primarily on the basis of whether the word denotes a material object (thus, primarily on the basis of visual and haptic experience). Connell and Lynott do not consider the possibility that internal information might play a role with respect to the dichotomy abstract versus concrete and therefore suggest developing a new measure of concreteness which can substitute for both the old measure of concreteness and that of imageability. This new measure should rely on instructions in which participants are explicitly asked to consider each of the five perceptual modalities in turn: auditory, gustatory, haptic, olfactory and visual.

Also on the basis of Connell's and Lynott's arguments, a new set of concreteness ratings was recently collected using different instructions which drive the participant to assign concreteness ratings considering all the external senses (Brysbaert et al. 2014). Even though they are quite long, the instructions of this new database need to be reported in full since they determine the specific idea of concreteness participants relied on to assign the ratings:

> Some words refer to things or actions in reality, which you can experience directly through one of the five senses. We call these words concrete words. Other words refer to meanings that cannot be experienced directly but which we know because the meanings can be defined by other words. These are abstract words. Still other words fall in-between the two extremes, because we can experience them to some extent and in addition we rely on language to understand them. We want you to indicate how concrete the meaning of each word is for you by using a 5-point rating scale going from abstract to concrete. A concrete word comes with a higher rating and refers to something that exists in reality; you can have immediate experience of it through your senses (smelling, tasting, touching, hearing, seeing) and the actions you do. The easiest way to explain a word is by pointing to it or by demonstrating it (e.g. to explain 'sweet' you could have someone eat sugar; to explain 'jump' you could simply jump up and down or show people a movie clip about someone jumping up and down; to explain 'couch', you could point to a couch or show a picture of a couch). An abstract word comes with a lower rating and refers to something you cannot experience directly through your senses or actions. Its meaning depends on language. The easiest way to explain it is by using other words (e.g. there is no simple way to demonstrate 'justice'; but we can explain the meaning of the word by using other words that capture parts of its meaning). Because we are collecting values for all the words in a dictionary (over 60 thousand in total), you will see that there are various types of words, even single letters. Always think of how concrete (experience based) the meaning of the word is to you. In all likelihood, you will encounter several words you do not know well enough to give a useful rating. This is informative to us too, as in our research we only want to use words known to people. We may also include one or two fake words which cannot be known by you. Please indicate when you don't know a word by using the letter N (or n). So, we ask you to use a 5-point rating scale going from abstract to concrete and to use the letter N when you do not know the word well enough to give an answer. (Brysbaert et al. 2014, p. 906)

These instructions rely explicitly on the standard picture of meaning according to which concrete terms are based on external perception while abstract terms rely on verbal information. Others than the instructions given by Spreen and Schulz (1966), these specify exactly how the contraposition between concrete and abstract should be understood, driving the participants to assign their ratings on the basis of the dichotomy perceivable through (at least one of) the senses versus linguistic construct.

The aim of the new instructions is to lead the participants to consider as concrete not only the terms denoting material objects (perceivable through the sight and through the touch), but also words whose denoting properties that are perceivable only through other senses (e.g. words denoting sounds, odors, tastes etc.). However, the analysis carried out by the authors on the ratings of the new database confirms the bias toward material objects: also with the new instructions people tended to assign a higher degree of concreteness to material objects:

> The high correlation between our ratings and those included in the MRC database (r = .92) attests to both the reliability and the validity of our ratings [...]. At the same time, the high correlation shows that the extra instructions we gave for the inclusion of nonvisual and action related experiences, did not seem to have much impact. Gustatory strength was not taken into account and auditory strength even correlated negatively, because words such as "deafening" and "noisy" got low concreteness ratings (1.41 and 1.69 respectively) but high auditory strength ratings (5.00 and 4.95). Apparently, raters cannot take into account several senses at the same time. (Brysbaert et al. 2014, p. 908)

Brysbaert and collaborators report that they found a "high correlation" between their ratings and those included in the MRC database. However, this overall correlation hides some differences that in our view are relevant. Indeed, if one considers specifically the words that are most problematic with respect to Paivio's theory and more generally with respect to the standard picture of meaning—i.e. those denoting x-ceptive states (i.e. affective, interoceptive and proprioceptive states)—and compares the concreteness ratings of these words in the MRC database and in Brysbaert and collaborators' collection, the result is quite different. Indeed, we confronted the concreteness ratings of the x-ceptive words selected for our previous study (Dellantonio et al. 2014) in the two databases and found that whereas their mean concreteness was 428 in the MRC database, it was 492 in the Brysbaert and colleagues' database, and that the 64 points of difference are statistically significant, $t(48) = 6.1$, $p < 0.001$. The fact that concreteness ratings for these words increased is extremely relevant since it results in the reduction of the difference between imageability and concreteness values which in our view indicate the link between a word and information of internal kind. The reason why the concreteness ratings of at least some relevant items like the x-ceptive words are higher in the new collection with respect to MRC lies in the instructions given to the participants and especially in two aspects of them.

First of all, these instructions lead participants to 'externalize' the criteria used to represent word meanings. The example of 'sweet' is highly significant in this respect: "The easiest way to explain a word is by pointing to it or by demonstrating it (e.g. to explain 'sweet' you could have someone eat sugar.)". 'Sweet' denotes a flavor. If we consider as concrete all words denoting something perceivable with one of the external senses, then the word 'sweet' is certainly concrete. However, its concreteness is not due to the fact that we can imagine the visual picture of somebody eating sugar, as suggested by the example, but it is due to the fact that we can directly perceive this sweetness through our sense of taste. If we observe somebody else eating something we never tasted before using our own papillae, we cannot know if it is sweet or not. In this sense, Brysbaert and collaborators' instructions are misleading since—in order to pursue the aim of making flavors appear more concrete to the participants and more close to the experience of material objects—they present the sensation of taste in terms of the visual perception of someone else eating. In this way, these instructions disregard the difference between internal (subjective) and external experience—i.e. between tasting 'sweet' directly and seeing somebody else eating something sweet—leading people to reinterpret their internal experience entirely in terms of external experience.[8] This is the reason why—following this instructions—people might have assigned a higher concreteness rating to all kind of internal sensations, interpreting interoceptive, exteroceptive or affective states more or less like tastes, sounds or smells. In the spirit of these instructions, since we can easily imagine the visual picture of someone moving, or making a suffering face because of the pain, or expressing his/her happiness by making a happy face, 'movement', 'pain' and 'happiness' must be considered as (fairly) concrete words. However, in this way the difference between concreteness and imageability disappears together with the difference between internal and external sensory experience.

Secondly, the instructions implicitly link concrete words with ostension, suggesting that a world is concrete if we can point to what it denotes. At the same time they suggest that the concreteness of actions should also be assessed this way: "to explain 'jump' you could simply jump up and down or show people a movie clip about someone jumping up and down". As in the case of 'sweet', also 'jump' leads to an externalized interpretation of the experience we use to understand words: we are not asked to think of the sensations we subjectively experience when we jump, but we are asked to take an external perspective and to picture someone jumping. In this way the participants are led to disregard the dynamical, first person information they have on the active performance of the action and to think of an action as they observe it externally when it is performed by someone else. Thus, instructions lead the participants to think of actions as if they were objects that can be perceived only through external observation. This externalization of actions produces the same effect discussed in

[8]The sense of taste belongs to the traditional five external senses, however, traditionally it is considered the most internal among the external senses in opposition to vision, which is considered as the external sense *par excellence*. While vision easily allows us to identify intersubjectively the source of the stimulation and is therefore considered to be also the most objective and real among the external senses, the sense of taste is the less objective because it is impossible to verify intersubjectively what a person is experiencing. For this reason, the notion of 'taste' has been generalized to indicate subjective judgments that cannot be disputed and are therefore potentially arbitrary (e.g. "it is a matter of taste"). For some philosophical reflections on the sense of taste in this vein see e.g. Arendt (1978, 1982).

the previous step: all words describing actions are evaluated as being concrete. Since their concreteness ratings are higher, the difference from their imageability becomes minimal and it becomes impossible to assess whether proprioceptive information plays a role with respect to the understanding of these words.

This externalization of the sensations and of actions makes this new database for concreteness (i) methodologically incommensurable with respect to MRC (and this in spite of the high correlation between the ratings reported in Brysbaert et al. (2014) and those included in the MRC database) and (ii) inadequate to assess a three-dimensional view of the difference between abstract and concrete in terms of which word meanings reflect different degrees of involvement of different kinds of external and internal sensory information as well as linguistic information.

As we showed above, the old concreteness and imageability scales can provide some useful indications to assess the different degrees of involvement of external, internal and linguistic information. However, they are still much too inaccurate and a more specific measure would be needed to assess abstractness and concreteness according to this new three-dimensional perspective. Ideally, such a measure should—among other things—include a distinction between grammatical categories (in particular between nouns and verbs) since on the basis of our discussion of the externalization of actions, one could hypothesize that verbs rely more heavily on internal (i.e. proprioceptive) information than the corresponding nouns, and specifically that e.g. 'to run' might evoke motor information on the performance of the action, while 'run' as a noun might evoke a more passive and externalized representation of a run.

5 Concluding Remarks

This study challenges the standard picture of meaning according to which words either rely on external sensory experience or are linguistic constructs and argues for the view that word meanings reflect different degrees of involvement of different kinds of external and internal sensory information as well as linguistic information. The paper addresses not only the theoretical aspects of this issue, but also its methodological consequences. In this regard, we specifically discuss whether the measures of concreteness and imageability used in the standard theory which opposed concrete and abstract words, might or might not continue to be useful since they allow us to assess the contribution of external and internal information. In the context of this discussion we suggest that, even though the classical constructs of concreteness and imageability might in fact be used as a joint measure to evaluate, at least approximately, whether a word representation relies on internal information, a more specific measure is needed for this aim.

References

Achinstein, P. (1965). The problem of theoretical terms. *American Philosophical Quarterly*, *2*(3), 193–203.

Ackrill, J. (1963). *Aristotle. Categories and de interpretatione*. Oxford: Clarendon Press.

Altarriba, J., & Bauer, L. (2004). The distinctiveness of emotion concepts: A comparison between emotion, abstract, and concrete words. *The American Journal of Psychology*, *117*(3), 389–410.

Altarriba, J., Bauer, L., & Benvenuto, C. (1999). Concreteness, context availability, and imageability ratings and word associations for abstract, concrete, and emotion words. *Behavior Research Methods*, *31*(4), 578–602.

Arendt, H. (1978). *The life of mind*. New York: Harcourt.

Arendt, H. (1982). *Lectures on Kant's political philosophy*. Chicago: Chicago University Press.

Barsalou, L. W. (1999). Perceptual symbol systems. *Behavioral and Brain Sciences*, *22*(4), 577–660.

Barsalou, L. W. (2008). Grounded cognition. *Annual Review of Psychology*, *59*, 617–645.

Barsalou, L. W., Niedenthal, P. M., Barbey, A. K., & Ruppert, J. A. (2003). Social embodiment. In B. H. Ross (Ed.), *The psychology of learning and motivation. Advances in research and theory*. Amsterdam: Academic Press.

Barsalou, L. W., & Wiemer-Hastings, K. (2005). Situating abstract concepts. In D. Pecher & R. Zwaan (Eds.), *Grounding cognition: The role of perception and action in memory, language, and thought* (pp. 129–163). New York: Cambridge University Press.

Berthoz, A. (2000). *The brain's sense of movement*. Cambridge: Harvard University Press.

Brown, D. J. (2006). *Descartes and the passionate mind*. Cambridge: Cambridge University Press.

Brysbaert, M., Warriner, A. B., & Kuperman, V. (2014). Concreteness ratings for 40 thousand generally known English word lemmas. *Behavior Research Methods*. 46(3), 904–911. doi:10.3758/s13428-013-1403-5.

Candlish, S., & Wrisley, G. (2012) "Private language", The Stanford encyclopedia of philosophy (Summer 2012 Edition), Edward N. Zalta (Ed.), http://plato.stanford.edu/archives/sum2012/entries/private-language/.

Carnap, R. (1956). The methodological character of theoretical concepts. In H. Feigl & M. Scriven (Eds.), *Minnesota studies in the philosophy of science I* (pp. 38–76). Minneapolis: University of Minnesota Press.

Coltheart, M. (1981). The MRC psycholinguistic database. *The Quarterly Journal of Experimental Psychology Section A: Human Experimental Psychology*, *33*(4), 497–505.

Connell, L., & Lynott, D. (2012). Strength of perceptual experience predicts word processing performance better than concreteness or imageability. *Cognition*, *125*(3), 452–465.

Cook, J. W. (1965). Wittgenstein on privacy. *The Philosophical Review*, *74*(3), 281–314.

Craig, A. D. (2003). Interoception: The sense of the physiological condition of the body. *Current Opinion in Neurobiology*, *13*(4), 500–505.

Craig, A. D. (2009). How do you feel-now? the anterior insula and human awareness. *Nature Reviews Neuroscience*, *10*(1), 59–70.

Craig, A. D. (2010). The sentient self. *Brain, Structure & Function*, *214*(5–6), 563–577.

Cruse, A. (2000). *Meaning in language. An introduction to semantics and pragmatics*. Oxford: Oxford University Press.

Damasio, A. (1999). *The feeling of what happens. Body and emotions in the making of consciousness*. Orlando: Harcourt.

Damasio, A. (2010). *Self comes to mind. Constructing the conscious brain*. New York: Pantheon.

Dellantonio, S., & Pastore, L. (2006). What do concepts consist of? The role of geometric and proprioceptive information in categorization. In P. Hanna, A. McEvoy, & P. Voutsina (Eds.), *An anthology of philosophical studies* (pp. 91–102). Athens: Athens Institute for Education and Research.

Dellantonio, S., Mulatti, C., Pastore, L., & Job, R. (2014). Measuring inconsistencies can lead you forward: Imageability and the x-ception theory. *Front Psychology*. doi:10.3389/fpsyg.2014.00708.

Dixon, T. (2003). *From passion to emotions. The creation of a secular psychological category*. Cambridge: Cambridge University Press.
Ekman, P. (1994). Moods, emotions and traits. In E. Ekman & R. Davidson (Eds.), *The nature of emotion: Fundamental questions* (pp. 56–58). Oxford: Oxford University Press.
Ekman, P. (1999). Basic emotions. In T. Dalgleish, & M. Power (Eds.), *Handbook of cognition and emotion* (pp. 45–60). Sussex, UK: Wiley.
Ekman, P., Sorenson, E., & Friesen, W. (1969). Pan-cultural elements in facial displays of emotion. *Science, 164*(3875), 86–88.
Gibbs, R. W., Jr. (1994). *The poetics of mind: Figurative thought, language, and understanding*. New York: Cambridge University Press.
Goldman, A. (1993). *Philosophical application of cognitive science*. Boulder: Westview Press.
Johnson, M. (1987) *The body in the mind: The bodily basis of meaning, imagination, and reason*. Chicago: University of Chicago Press.
Kassam, K., Markey, A., Cherkassy, V., Loewenstein, G., & Just, M. (2013). Identifying emotions on the basis of neural activation. *PLoS ONE, 8*(6), e66032.
Knorrp, W. M. (2003). How to talk to yourself or Kripke's Wittgenstein solitary language argument and why it fails. *Pacific Philosophical Quarterly, 84*(3), 215–248.
Kousta, S., Vigliocco, G., Vinson, D., Andrew, A., & Del Campo, E. (2011). The representation of abstract words: Why emotion matters. *Journal of Experimental Psychology: General, 140*(1), 14–34.
Kripke, S. (1982). *Wittgenstein on rules and private language*. Oxford: Blackwell.
Lakoff, G. (1987) *Women, fire, and dangerous things: What categories reveal about the mind*. Chicago: University of Chicago Press.
Lakoff, G. & Johnson, M. (1980) *Metaphors we live by*. Chicago: University of Chicago Press.
Lakoff, G. & Turner, M. (1989) *More than cool reason: A field guide to poetic metaphor*. Chicago: University of Chicago Press.
Lebuffe, M. (2010). The anatomy of the passions. In O. Koistinen (Ed.), *The Cambridge companion to Spinoza's ethics* (pp. 188–222). Cambridge: Cambridge University Press.
Nigel, T. (2013). Mental imagery. In E. Zalta (Ed.), *The Stanford encyclopedia of philosophy*. http://plato.stanford.edu/archives/fall2013/entries/mentalimagery/.
Paivio, A. (1965). Abstractness, imagery, and meaningfulness in paired-associate learning. *Journal of Verbal Learning and Verbal Behaviour, 4*(1), 32–38.
Paivio, A. (1986). *Mental representations. A dual coding approach*. Oxford: Oxford University Press.
Paivio, A. (2007). *Mind and its evolution: A dual coding theoretical approach*. Mahwah: Erlbaum.
Paivio, A., Yullie, J., & Madigan, S. (1968). Concreteness, imagery, and meaningfulness values for 925 nouns. *Journal of Experimental Psychology. Monograph Supplement, 76*(1), 1–25.
Papineau, D. (1996). Theory-dependent terms. *Philosophy of Science, 63*(1), 1–20.
Plutchik, R. (1980). A general psychoevolutionary theory of emotion. In R. Plutchik & H. Kellerman (Eds.), *Emotion: Theory, research, and experience: Vol. 1. Theories of emotion* (pp. 3–33). Academic Press: New York.
Prinz, J. (2002). *Furnishing the mind: Concepts and their perceptual basis*. Cambridge: MIT Press.
Prinz, J. (2004). *Gut reactions. A perceptual theory of emotion*. Oxford: Oxford University Press.
Reizenzein, R. (2009). Emotional experience in the computational belief-desire theory of emotion. *Emotion Review, 1*(3), 214–222.
Schroeder, S. (1998). *Das Privatsprachen-Argument. Wittgenstein über Empfindung und Ausdruck*. Paderborn: F. SchöningVerlag.
Spreen, O., & Schulz, R. (1966). Parameters of abstractions, meaningfulness and pronunciability. *Journal of verbal learning and Verbal Behaviour, 5*(5), 459–468.
Tomkins, S. (1962). *Affect, imagery, consciousness: Vol. I: The positive affects*. New York: Springer.
Tomkins, S. (1963). *Affect, imagery, consciousness: Vol. II: The negative affects*. New York: Springer.

Vigliocco, G., Meteyard, L., Andrews, M., Kousta, S. (2009). Toward a theory of semantic representation. *Language and Cognition, 1*(2), 219–247.

Wilson, M. (1988). The MRC psycholinguistic database: Machine readable dictionary, version 2. *Behavioural Research Methods, Instruments and Computers, 20*(1), 6–11.

Wittgenstein, L. (1953). *Philosophical investigations*, translated by G. E. M. Anscombe, Oxford: Blackwell (3rd ed.) 1967.

An Eco-Cognitive Model of Ignorance Immunization

Selene Arfini and Lorenzo Magnani

Abstract In 2005, Woods described the epistemic bubble as an immunized state of human cognition that compromises the awareness of the agent about her beliefs and knowledge. The idea of an immunized knower swung with the proposal advanced by Gabbay and Woods of constructing a practical logic and epistemology, which can actually define itself as agent-centered, goal-oriented, and resource-bound. In order to carry out this project, in this paper we will introduce a symmetrical view on the agent immunization, focused on the agent's missing awareness of her ignorance, also highlighting the importance of considering the actual agent as cogently ignorant, too. Eventually, we will formulate an idea of ignorance that can be fruitfully studied in the newborn field of naturalized epistemology and logic.

1 Introduction

The cognitive traits and dynamics that alter and compromise human rationality have been the ambitious core of numerous philosophical and cognitive-oriented research in the last three decades. A recent project has surveyed this topic (Gabbay and Woods 2003, 2005; Woods 2013), aimed at drawing a logical model of actual human cognition and rationality. In the perspective of that project, the intent is to put our attention on the generation of ignorance beyond the boundaries of the analysis of fallacious reasoning and heuristic strategies. Woods (2005) introduced the notion of epistemic-bubble as an "epistemic immunization": we mean to introduce a symmetrical "ignorance immunization", in the first-person perspective of the actual agent.

S. Arfini (✉)
Department of Philosophy, Education and Economical-Quantitative Sciences,
University of Chieti and Pescara, Chieti-Pescara, Italy
e-mail: selene.arfini@gmail.com

L. Magnani
Department of Humanities, Philosophy Section and Computational Philosophy Laboratory,
University of Pavia, Pavia, Italy
e-mail: lmagnani@unipv.it

L. Magnani et al. (eds.), *Philosophy and Cognitive Science II*,
Studies in Applied Philosophy, Epistemology and Rational Ethics 20,
DOI 10.1007/978-3-319-18479-1_4

We will start by analyzing the problematic introduction of the problem of ignorance in the system of logic and epistemology focused on actual human cognition and rationality proposed by Gabbay and Woods (Sect. 2), enhancing the Fallibilist principles of the New Logic also adding (Sect. 3) a Negative affirmation about the ability of the agent to recognize her own ignorance. Then, we will proceed by illustrating the Epistemic Bubble Thesis (Woods 2005) in order to show an ignorance-sensitive perspective on the epistemological immunization of the agent (Sect. 4) and we will see how the defeasible mechanism of ignorance-detection also affects knowledge-recognition. Finally, we will define a model related to the ways an agent adopts in order to manage the autoimmune structure of knowledge and beliefs (a sort of useful Homunculus Fallacy), the consequences of which can be usefully studied in the newborn field of naturalized logic (Sect. 5).

2 Introducing Ignorance into the Naturalization of Logic

In 2003 Gabbay and Woods officially proposed a program, condensed in a series of volumes called "A Practical Logic of Cognitive Systems", to the aim of constructing new logical models able to fill the gap between the logical and cognitive representation of human agent and its "real" —eco-cognitive multi-dimensioned—counterpart. The last volume of the series, published in 2013, was aimed at drawing an empirically "sensitive" and "aware" form of logic, able to deal with actual reasoners' cognitive performances (Woods 2013).[1] At the same time the volume is a collection of the logical and cognitive studies concerning errors in reasoning and their productive character.[2]

In this massive production, even if the focus has been on the third-way reasoning[3] humans actually performs, especially the exploitation of errors and downfalls, the agent is always considered an enough acquainted reasoner, a *knower*, in her intents. The principles of her possibilities and boundaries are determined by general abundance theses that substantiate a form of fallibilism:

Proposition 3.2b

THE COGNITIVE ABUNDANCE THESIS: Human beings have knowledge, lots of it.

[1]Woods' "empirically sensitive" logic refers to the construction of logical systems able to take advantage of the results of cognitive science and its empirical results. The relevance of the "empirically aware" character refers to the importance attributed to the study of actual human cognition. As they say by "adjusting its [the logic] provisions to the cognitive natures of real life reasoning agents" (Woods 2013).

[2]Cf. the other volumes of the series (Gabbay and Woods 2003, 2005).

[3]Third way reasoning is "the reasoning which, when good, is made so by circumstances other than deductive validity or inductive strength" (Woods 2013, p. 32) humans actually performs.

Proposition 3.2c

THE ERROR ABUNDANCE THESIS: Human beings make errors, lots of them.

Proposition 3.2d

THE ENOUGH ALREADY THESIS. Human beings are right enough about enough of the right things enough of the time to survive and prosper (and occasionally build great civilizations) (Woods 2013, pp. 86–88).

In this perspective the presentation of the "right enough human being" is referred to the study of the cognitive endowments of an "actual" agent, mostly focusing on her knowledge and cognitive skills. The "ignorant" part is basically described as an innocent tendency to commit errors (even if lots of them) or treated in the light of fallacious reasoning. We contend, following Proctor, that "ignorance is more than a void" (Proctor 2005, p. 2), it is an influential part of human cognition, and affects not only our deficiencies but also the ways we adopt to fill them with beliefs and knowledge. Certainly fallacies and heuristics have always been considered the main door to step into the problem of "ignorance"[4] and other research on less deceptive inferential activities (even prompted to promote the Naturalization of Logic[5]), have stressed the so-called ignorance-preserving traits of abduction.[6] However, aiming at furnishing a new contribution to the ambitious project of the naturalization of logic, we will introduce an explanation of the role played by the ignorant part of the "real agent". First of all, by analyzing the state that most indicates the presence of ignorance in the perspective of the agent herself, that is "the state of doubt" and, second, by showing how this state affects the Fallibilist principles (which are the base of the Abundance and the Enough Already Thesis) proposed by Woods.

2.1 *The Visible Part of Ignorance: Peirce's Irritation of Doubt*

Despite the topic of "doubt" undeniably holds a rich past in the history of philosophy, the interest around it in the last century has progressively decreased mainly because of the focus of analytical philosophy on the definitions of knowledge and truth. Many authors became more interested in specifying the visible boundaries that characterize certainty than in directly examining the nucleus of what is beyond it. Attention has been devoted to intertwine doubt with specific arguments such as ambiguity, vagueness, and credibility.[7]

---

[4]See, for instance the classical (Hamblin 1970; Walton 1995; Woods et al. 2004).

[5]Cf. Magnani (2015).

[6]Cf. Aliseda (2005), Magnani (2013), and Gabbay and Woods (2005).

[7]Ambiguity, vagueness, and credibility in the field of informal logic and critical thinking are illustrated in the last edition of *Critical Thinkings*, by Moore and Parker (2012) and in a seminal article

On the contrary, the philosophical background that informs the Naturalization of Logic resorts to Peirce and Peircean tradition: Peirce directly examined the problem of doubt and tried to grasp its philosophical, epistemological, and cognitive essence. Peirce's exposition of the dynamic between doubt and belief [(Peirce 1931–1958, Book III, Chap. 4), (Peirce 1992–1998, Chaps. 7 and 8)] is focused on the fundamental gap between ignorance and knowledge in terms of what is and what is not (or it never will be) believed. In this framework, the epistemic status of doubt is seen as the conscious experience of a missing answer for a problem. The agent in the state of doubt cannot proceed to act and consequently gets frustrated.

> Doubt and Belief, as the words are commonly employed, relate to religious or other grave discussions. But here I use them to designate the starting of any question, no matter how small or how great, and the resolution of it. [...] Most frequently doubts arise from some indecision, however momentary, in our action. Sometimes it is not so. [...] However the doubt may originate, it stimulates the mind to an activity, which may be slight or energetic, calm or turbulent. Images pass rapidly through consciousness, one incessantly melting into another, until at last, when all is over – it may be in a fraction of a second, in an hour, or after long years – we find ourselves decided as to how we should *act* under such circumstances as those which occasioned our hesitation. In other words, we have attained belief (Peirce 1998a, pp. 127–128).

In Peirce's pragmatist theory, the specific difference between the two states of mind—as we can read in the above passage—is practical. He pictured the transition between the state of doubt to belief in terms of *action* and *reaction* of the agent who feels them. The relationship of the mental state of doubt with the active start of questioning and of the state of belief with the relief of the discovery of an answer is fundamental. It allows to clarify the profound connection between the epistemic conditions of the agent and her cognitive *reaction* to them. Obviously, the reasons to get a reliable answer to the problems that trigger the state of doubt could be many. The agent may desire to attain some information to overtake a quick moment of indecision so claiming further investigation in order to solve a "great question". However, according to Peirce, the main incentive that drives the agent to find a solution of the problems that torment her is the cognitive and psychological state related to the doubt itself: in particular, the *known difference between the feelings* that doubt and belief provoke. In another famous article, Peirce described the states of doubt and belief as antithetical (Peirce 1998b), precisely in consideration of this aspect. Belief is considered the quiet state of affirming a principle (a proposition, an idea) and doubt an irritating condition, which not only deprives the agent of her certainties but, through that loss, compromises her quiet.

In this sense, what determines Peirce's definition of doubt and belief is the diversity of feelings and behaviors they generate. The peaceful state of belief prompts—through either the adoption or the defense of a principle (a proposition, an idea)—an agent to act. It creates a state of satisfaction on the agent's mind who is ready to perform various actions based on her confidence in her beliefs. On the contrary,

(Footnote 7 continued)
about the distinction between denotational ambiguity and vagueness, (Dunbar 2008). Cf. also the classical work of Grice on *implicatures* (Grice 1975).

doubt is characterized by a state of frustration caused by either the lack of knowledge or the falsification of a principle previously adopted. Indeed, the laborious work of investigation triggered by doubt, and portrayed with some emphatic words by Peirce, can be described as the position of a very specific question that is only raised by (but not ends with) the negation of a precedent, unconfirmed, belief. It is a state of agitated research and craving for an answer. It is also metaphorically described as the irritation of a nerve, in a text that is worth to be quoted:

> Thus, both doubt and belief have positive effects upon us, though very different ones. Belief does not make us act at once, but puts us into such a condition that we shall behave in some certain way, when the occasion arises. Doubt has not the least such active effect, but stimulates us to inquiry until it is destroyed. This reminds us of the irritation of a nerve and the reflex action produced thereby; while for the analogue of belief, in the nervous system, we must look to what are called nervous associations – for example, to that habit of the nerves in consequence of which the smell of a peach will make the mouth water (Peirce 1998b, p.114).

What Peirce defines as the *irritation of doubt* is an unwanted state of mind caused by the loss of certainty in the agent knowledge. We can consider this description as the easier way to see the *experience* of a part of ignorance. The individual desperately wants to escape from the condition of doubt (Peirce 1998b) because, if belief is (at least) the confidence about having a reliable knowledge in order to act, the state of doubt implies the possibility of a blind spot in that knowledge, a missing direction to move toward. Ignorance, in the most visible and concrete form, appears to be just the formulation of specific doubts.

In order to complete this consideration we should mention that it is obvious that we do not refer to doubt as a skeptical form of abyssal negation. As repeatedly affirmed, our analysis is framed by the Actually Happens Rule (Gabbay and Woods 2001), which determines the cognitive target of the Naturalization of Logic[8]: the state of doubt is a state that an actual agent can experience every day. Often, it can be devised through a directed question which exhibit a specific (broader or less) blind spot recognized by the individual.

Following a more classical definition of doubt, in Hegel's terms we can say that it is a "determinate doubt", which has an epistemic state and a specific content. "This is just the Skepticism, which only ever sees is *pure nothingness* in its result and abstract from the fact that *this nothingness is specifically the nothingness of that from which it results*. For it is only when it is taken as the result of that from which it emerges that it is, in fact, the true result: in that case it is itself a *determinate nothingness*, one which has a *content*" (Hegel 1977, p. 51).

In summary, our introduction to the problem of ignorance in the Naturalization of Logic clearly relates to the definition of doubt provided by Peirce, which appears to play the conceptual role of a perfect medium term between ignorance and knowl-

[8]The Actually Happens Rule suggests to investigate the epistemic status of an ordinary agent instead of studying the impeccability of an ideal one—hence studying what *actually happens*. To be precise, the rule claims: "To see what agents should do, look first to what they actually do. Then repair the account if there is particular reason to do so" (Woods 2005, p. 734).

edge. For this reason, we can take advantage of this Peircean definition to insert a complementary proposition into the Fallibilist principles we mentioned above.

3 Fallibilism: A Belief-Based Paradigm

In order to comprehend how Peirce's epistemology grounds the Naturalization of Logic we should leave for a moment the analysis of doubt and briefly revisit the definition of belief. Peirce describes the state of belief as having just three properties: "first, it is something that we are aware of; second, it appeases the irritation of doubt; and, third, it involves the establishment in our nature of a rule of action, or, say for short a *habit*" (Peirce 1931–1958, 5.397). Thanks to this scheme rejoin and further deepen the main tenets of Woods' fallibilism indicated in the previous subsection.

The awareness of our belief state (that is, according to both Peirce and Woods, the only state that allows us "to know")[9] obviously is what makes us able to define ourselves "knowers". In this perspective, the tendency to knowledge indicated by Woods in the proposition 3.2b is just a consequence of the awareness of how much we believe we know and how much we are able to learn.

The second feature, the capacity of belief to appease "the irritation of doubt" is at the basis of the "Error Abundance" thesis, which composes the second item of Woods' fallibilism. Believing is a *pleasurable* state, a state that calms the agent and gives her the cognitive resources to act. It is this practical advantage that makes it preferable to doubt, no matter if it is *epistemically* more convenient or less. The tendency to commit errors (and so of believing in an incorrect, or "fast and frugal", statement instead of doubting it)[10] of the actual agent is exactly derived from this unfortunate preference.

Finally, the third condition of belief, which "involves the establishment in our nature of a rule of action, or, say for short a *habit*", can be seen as the feature that seals the "Enough Already Thesis". Even intuitively, believing to know something has two main consequences: (a) it repels the irritation of doubt, making us sure about our own knowledge (sometimes compromising our ability to individuate errors in it) and, (b) since belief gives us the possibility to act in the world upon a certain circumstance, we will be inclined to rely on the same belief as a principle for solving other similar circumstances. The "enough already thesis" does not affirm much more than the prevalence of the occurrence for our belief to be confirmed by a personal (more or less fortunate) experience.

---

[9]Speaking of belief as the condition for knowledge we do not assume that there is no possibility of having a form of knowledge that the agent is not fully aware of [as Polanyi named it, a "tacit" form of knowledge (Polanyi 1966)]. Simply, in order to deal with the Naturalization of Logic and its paradigms, we prefer to use a stronger meaning than a broad sense of knowledge, speaking of it as the conscious attainment of reliable sentences or propositions.

[10]Here, thanks to Gigerenzer's formula, (Gigerenzer et al. 1999), we are in general referring to the cognitive virtues of heuristic reasoning and fallacies, analyzed by informal logic, psychology, and cognitive science in the past forty years, cf., for example, (Gigerenzer and Goldstein 1996; Woods 2007; Ippoliti 2015; Magnani 2014).

3.1 Another Kind of Abundance: How Fallibilism Can Shape Ignorance

The parallel between Woods' Fallibilist principles and the Peircean definition of belief allows us, first of all, to confirm the knowledge-based perspective of Woods' analysis itself. The actual agent is a *knower*, because she is also a *believer*. The fact she believes she knows implies the possibility of committing errors, but it does not compromise her epistemic status of knower. We contend, extending Woods' characterization of the actual agent, that ignorance is more than a simple tendency to commit errors: taking advantage of the description of belief in Peirce's work we can put down the equivalent three properties for doubt considering it as visible ignorance.

1. Like belief, doubt is a state *we are well aware of*;
2. it is an *unwanted* and *irritating* state for the agent;
3. it requires an *inferential reasoning* (and the fixation of another belief) in order to end.

Two conclusions follow, one more evident than the other.

The most evident outcome is a definition of ignorance that is already formulated in Woods' theory. He defines ignorance as "inferentially productive", as a part of our cognition that we can examine through fallacious but effective inferential processes (Woods 2013, p. 335). In this perspective we can rethink the principles of knowledge and error abundance adding a "doubt openness condition". The possibility of doubt and of recognizing and admitting ignorance opens the possibility of an improvement of the agent knowledge and so it enforces the tendency to gain new data (Knowledge Abundance). At the same time, as already said, doubt also implies a cognitive irritation that forces the agent to quickly arrive to a resolution of the problem at stake. This urgency can affect the inference, performed in order to solve the problem, making easy for the agent to commit errors (Errors Abundance).

The second and less evident consequence of introducing the doubt as "the visible part of our ignorance" in the fallibilist triad, is instead a sort of "negative" affirmation. Examining the epistemic status of ignorance at the conscious level, so addressing knowledge firstly as belief, and speaking of doubt as something that we do not recognize as belonging to our knowledge, we let the door open to the fact that an actual agent is not "simply" ignorant of what she is aware she doesn't know. Ignorance is not completely equivalent to doubt, it is not just a missing piece of our cognition, something that the agent knows she does not know. Doubt can be *perceived* as the frame of our ignorance, but it is the simple consequence of our fallible cognition. Of course the project toward the Naturalization of Logic has already proposed to extend the field in which knowledge can be investigated, contending the importance of the examination of the errors of reasoning and their "positive" aspects. Something more can be said: since ignorance corresponds to something that goes, at least partially, beyond the cognitive sight of the agent, we should investigate it as the Naturalization of Logic aims at explaining the limit of our knowledge: far beyond the self-perspective of the actual agent.

The "negative" affirmation permits us to have an "Enough Already Thesis", less indulgent with respect to the agent's actual status. As we have already said, the psychological and emotional component of doubt makes its experience repulsive for the agent. So, if even the visible part of ignorance is hard to be managed by the agent, the part of ignorance that falls beyond her control (or her will) must be extremely difficult to reach. The "Enough Already Thesis" displays the capacity of human beings to be right enough about enough of the right things enough of the time to survive and prosper (Woods 2013, pp. 86–88). Now we should add: "despite" *how much the agent ignores*, how much *she does not want to admit she ignores* and the *repulsion for being in a state of doubt* the agent appears to be not just "able enough" to survive and prosper, but also to bear the weight of her ignorance without feeling it.

At this point, it is important to say that our goal is not simply to deal with the mere capacity of the actual agent to copy with situation of unknown reliability of her information. This topic has been examined through the last decades by numerous philosophers and theorists of various field of expertise. The theory of *Bounded Rationality* is the most popular result of this investigation. Herbert Simon—who put on the scenes the idea of a bounded rationality (eco-logically constrained) (Simon 1993, 1997)—illustrated that humans can make decisions and solve problems in presence of uncertainty, incompleteness, and unreliability of the information they possess. This trait is considered the reason that permits the actual agent to revise the results of her inferences once she finds additional significant information (Simon 1997). Although it is an interesting point of investigation, this intellectual tradition is restricted to display the capacity of human beings to manage eco-cognitive bounded *knowledge*. In order to continue our analysis on ignorance, we instead have to change the focus and to give heed to what *it is not at the hand* of the agent, her unseen chances, beyond the limits she knows.

Hence, instead of focusing on the confidence in the "Enough Already Thesis" (that we can condense in "we are able to survive, after all"), all the added *caveats* demand a deeper questioning on the tendency of human beings to avoid a complete awareness of their own ignorance. It is not unreasonable asking how and why this is functional, for instance. The examination of these important issues is already displayed in what Woods called the Epistemic Bubble (Woods 2005). We can picture the Epistemic Bubble as a form of knowledge-based immunization that inhibits the agent from distinguishing her knowledge and her beliefs. In the next section, we will investigate the "Bubble Thesis": this will allow us to comprehend that the agent bears also a *ignorance-based immunization*, which compromises her ability to frame her own ignorance and distinguish it from what she just doubts about.

4 The Bubble Thesis and the Double-Sided Autoimmunity System

The idea of the Epistemic Bubble originates from the analysis of both purpose and ending of the state of doubt, albeit it remains focused on the analysis of the state of belief. Citing Peirce's words, if "the production of a belief is the sole function of thought" (Peirce 1998a, p. 127), the research of new data and the inferential reasoning would stop when the agent builds up a belief, albeit *not necessarily the correct one*. The irritating character of the state of doubt implies that the agent can consider herself satisfied when she achieves a belief only "deemed" as trustful, and not "undeniably" so. The state of belief is not just pleasant, but also fallible and uncertain. In its essence, Woods' Bubble Thesis focuses on the relation between the complex of beliefs an agent has and her awareness as regard as either their correctness or unsteadiness. This suggests that the agent's mechanism of belief formation can provide an easy way out to the Peircean irritation of doubt through a systematic *ascription of knowledge* concerning a mere belief, immunizing the agent from being able to spot the difference (so letting the agent think she knows something when she merely believes she does).

In order to utterly understand the potentialities of this idea, we have to introduce two dichotomies that Woods indicates as substantial. The first between what we could consider a broad definition of knowledge related to the Peircean state of belief. The second concerning the important distinction between the *first* and *third-person* perspective of the agent.

4.1 The Downside of Belief and of the First-Person Perspective

As already mentioned, belief in Woods' theory corresponds to the Peircean definition: it is the sole state that solves the irritation of doubt and brings peace to the cognitive unsteadiness of the agent; fundamentally, it is a state that calms the agent's mind. Knowledge, instead, is defined as a "*kind of case-making. One knows that P only if one has one's disposal a case of requisite strength to make for P*" (Woods 2005, p. 735). The distinction between belief and knowledge, however, is not evident for the agent who knows and believes. Indeed, the achievement of knowledge always entails a state of belief in the agent, even if the attainment of a belief does not directly imply the gain of knowledge.

The entanglement between knowledge and belief drives our argumentation to the difference between the first and the third-person perspective. Indeed, for the agent is kind of easy to say if someone else *knows* or *thinks she knows* something. That is to say, from the third-person perspective one can tell the difference between a belief that stands for an actual knowledge attainment and a belief that just brings about some cognitive relief to an irritating state of doubt. The agent can judge if someone else's is either effective knowledge or mere confidence. From the first-person perspective, the difference is instead blurred, due to the fact a belief state entails the occurrence of knowledge. This is an entanglement indeed recognized as the focus for the asymmetry between first and third-person perspective. Whenever the agent knows something, *she is compelled to believe she knows it*. But, since the attainment of knowledge is different from the establishment of a belief, she can believe she knows something even when she does not. This distinction between knowledge and its mere ascription is visible only in a third-person perspective. As reported in the Proposition 4 "(Belief as knowledge-ascription). Whenever it is true for *Y* to say of *X* that *X* believes that *P*, it is also true that X takes himself as knowing that *P*" (Woods 2005).

Hence, while in the first-person perspective a reliable belief is always claimed as knowledge, in the third-person perspective the proposition can be judged as potentially verified or erroneous. Thus, in the case of the first-person perspective there is not a clear distinction between knowing and believing in something, even if it is pretty clear in the case of the agent's third-person perspective. At the same time, in the case of the first-person perspective the state of belief represents not only the way the agent can experience some relief from the irritation of doubt, but also the unique possibility for the agent of attaining any sort of knowledge. This idea is better expressed in Woods terms in the Proposition 6:

> *(The Downside of Belief). Belief is both a condition of knowledge and an impediment to its attainment.*
>
> In so saying, we can see that the traditional approach to knowledge is defective. It rightly insists on the indispensability of belief for knowledge, but it ignores, or downplays, its impedimental role. If this is right, then the capacity for, indeed the likelihood of, false apparency is structured by the phenomenology of cognitive states and reinforced by one's auto-psychology (Woods 2005, p. 739).

So, albeit the fact that *there is a solid difference* between the epistemological status of belief and knowledge, the agent cannot be aware of this distinction when she has to deal with her own cognition. Hence, in its essence, the *epistemic bubble* is configured as a *first-person knowledge-ascription*, performed by the knowing agent, to whom the difference between *knowing* something and *thinking she knows* that same thing is unapparent—and the tension that may arise is always solved in favor of the former (Woods 2005). This mechanism always provides—more or less heavily—an illusion about the truthfulness of the knowledge of the agent's first-person perspective.

Woods describes the epistemic bubble as an *autoimmune* mechanism of the agent. The naiveness of the agent about her own cognition is directed by the same system that at the same time permits her to attain any type of knowledge. Belief, as a cognitive structure, is *in primis* a tool that gives her the possibility of taking action into the world. If the agent could not be sure about what she thinks she knows, she could not take any decision and she would be constantly in a state of doubt and struggle. The autoimmune mechanism helps her out from the freezing state of doubt but does not provide a *safe* exit from it. Interestingly, Magnani (2011) argued that this mechanism, analyzed by Woods as far as a propositional/sentential kind of knowledge is concerned, may actually be extended to any kind of belief entertained by a subject (not necessarily expressed, or expressible, by language) to the point of illuminating a kind of wide *cognitive bubble* (also including bubbles that are potentially sharable such as the "moral bubble" and the "religious bubble").[11]

---

[11]The broad architecture of the *cognitive bubble* defined by Magnani (2011) frames how, in certain respects, human cognitive mechanisms need develop some autoimmune devices, or become to some extents self-blind. The "moral bubble" (Magnani 2011) captures how people, in order to engage any kind of moral behavior (typically involving punishment), must become blind and *autoimmune* to the possible violence they perform. While the epistemic bubble is typically conceived as a cognitive

As we can see, the epistemic bubble as an autoimmune mechanism concerns the limits of the attainment of knowledge, its entanglement with the state of belief, and the unapparent distinction between the two in the first-person perspective. As we have said in the case of the original Fallibilist principles mentioned in the previous section, the idea of epistemic bubble is profoundly connected with the definition of belief offered by Peirce. Using a similar connection, in order to shift the focus on the limits of ignorance-recognition, we must reconsider Peirce's doubt in the light of the autoimmune mechanism described above.

4.2 *Doubt and the Missing-Ascription of Ignorance*

Given the fact belief and knowledge are connected in the first-person perspective, but way far from each other in the third-person view, we can formulate the same consideration in the case of doubt and ignorance. The "negative" affirmation in the Fallibilist principles is oriented to highlight the distinction between doubt and ignorance, but this separation is manifest only in a third-person perspective. In the third-person perspective, doubt presents the character of being a state of irritation for the subject, a push for inferential reasoning, and, mainly, a state *she is aware of*: it is a frame of the ignorance of the subject *in those limits*. The proper ignorance of the agent is beyond the frame of her doubts. It is something the agent cannot consider in first-person perspective. At the same time, the only "visible part of ignorance" for the first-person perspective can entirely frame the ignorance of the agent.

This relation is clear when we think about the possibility of describing how we ignore something. The only method that we can apply is to frame our ignorance, speaking about the propositions we doubt to be true, the situations we are not certain about, and the collection of data we are not sure if they are reliable or not. But these data are just what *we consider* part of our ignorance. They cannot be even close to the propositions we are not informed of, the situations out of our sight, and the collection of data we are not aware of. These data are part of our ignorance, but we cannot reach them through our doubts. At the same time, doubt is the only cognitive tool that permits us to grasp pieces of ignorance and let us admit that there is something out of our reach.

So, exactly as in the epistemic bubble, albeit the fact that *there is a solid difference* between the epistemological statuses of doubt and ignorance, the agent cannot be aware of this distinction when she has to deal with her own cognition. Consequently,

(Footnote 11 continued)
constraint of the single individual as it portrays a single agent's cognitive structure, other kinds of embubblement are more or less prone to be culturally shared. The "religious bubble"—investigated in Magnani and Bertolotti (2011)—describes how the typical cognitive praxis of religious beliefs involves their enactment in certain social situations, for instance moral, spiritual, rhetorical, but their are deactivated when other kinds of decision or expectations are at stake (e.g. practical expectations in hunting, administering one's resources, and so on).

we can describe the *ignorance-based bubble* as a *missing-ascription of ignorance*, performed by the agent, to whom the difference between *ignoring* something and *doubting* is unapparent.

This structure is also an *autoimmune* mechanism of the agent. Doubt, the only tool that permits the agent to investigate a part of her ignorance, makes also *impossible* for the agent to distinguish the amount of actual ignorance she possesses from what she is just able to recognize. At the same time, without this autoimmune system we would never leave the state of doubt. Even if it was possible to think about our own ignorance in its entireness and deepness, it would have been completely disadvantageous! There is a quite substantial convenience to act without the complete awareness of our limits, as proven by the numerous studies we have quoted in the first section, on the cognitive virtues of fallacies and on "fast and frugal" heuristics.[12] But the consideration regarding the immunity of the agent about her own ignorance does not only concern the so-called "errors of reasoning". It is not a mechanism that involves the improvement of decision-making strategies with little loss of certainty. For instance, it does not coincides with Gigerenzer's "Law of Indispensable Ignorance" (Gigerenzer 2004), which describes the efficiency of the agent in a situation of bounded rationality (so in a condition of weak knowledge). The ignorance-based autoimmune mechanism illustrates the *ignorance about one's own ignorance* as the only possible condition for the attainment of any kind of knowledge in more or less uneasy condition. The immunization to ignorance is an indefeasible mechanism of human cognition as well it is the epistemic bubble. They simply define the borders of possibility for first-person perspective agents to modify their own epistemological status.

By considering the cognitive state of doubt, we can extend our analysis also considering Woods' thesis about *truth*. As we will better illustrate, the analysis of the epistemic bubble leads to the affirmation that the truth, for the first-person perspective, is a *fugitive* property. In brief, the difficulty for the agent to distinguish the difference between what she knows and what she believes, impairs her possibility to reach and

[12]The advantages of being unaware of the fallibility of our cognition is also recalled in the Proposition 6.1a and Corollary, and 6.1b, in Woods (2013): "[Proposition 6.1.a] It is sometimes reasonable to use procedures that lead to error. Blanket error avoidance is not, therefore, a general condition on cognitive success. [...] [Corollary] There is cognitive good to be achieved by the engagement of cognitive procedures that let us down with notable frequency. Such letdowns are occasion to learn from experience. They are fruitful contexts for trial by error. [...] [Proposition 61b] By and large, individuals have speedy and reliable feedback mechanisms" (Woods 2013, p. 185). The employment of error-permitting heuristics, especially in situations of cognitive economy (that is when the production and distribution of knowledge in the agent's environment is subject to some constraints—for short, everyday situations), renders the agent able (1) to *try* different patterns of reasoning in problem-solving processes and, (2) to *learn* from mistakes if and when they occur. Hence, since these useful heuristics are often fast to adopt and the errors easy to spot, they provide a clear knowledge-enhancement effect. Moreover, the exploitation of error-correcting and damage-managing strategies is considered cheaper and more productive in the temporal extended dimension than the adoption of totally error-free methods.

recognize truth. Using the same association in the case of the analysis of the agent's immunity to her own ignorance we can arrive to a similar consideration regarding her capacity to reach and recognize the entireness of ignorance beyond the frame of her doubts.

5 The Fugitivity of Truth (and Ignorance)

The autoimmune system of the epistemic bubble makes the attainment of truth a relatively impossible task from the first-person perspective, adding a veil of skepticism to the cognitive analysis. This is clearly stated in Proposition 15:

> PROPOSITION 15 (Fugitivity of truth). Within epistemic bubbles, truth is a fugitive property. That is, one can never attain it without thinking that one has done so; but thinking that one has attained it is not attaining it (Woods 2005, p. 745).

At this point it is interesting to note we can apply a similar argument when considering ignorance. The missing-ignorance ascription in the first-person perspective makes the idea of ignorance a "fugitive property" because every time the agent tries to define what she ignores, she is reaching just the limits of her doubts.

Hence, if we can describe the mechanism of epistemic embubblement taking advantage of a two-sided definition:

- the impossibility—from the first-person perspective—of a clear distinction between knowledge and belief,
- and the certainty of the agent to have a fully achieved knowledge about something even without the actual attainment of it,

we can find a similar two-sided definition in the case of the ignorance-based bubble:

- the impossibility—for the first-person perspective—of a clear distinction between doubt and ignorance,
- and the certainty of the agent to have fully framed her ignorance trough her doubt, even if she cannot do it.

As we have already mentioned, the ignorance that the agent can perceive is just defined through her doubts, and her doubts can picture just a small portion of her ignorance. The disparity between the two parts of her ignorance can be illustrated using the Freudian metaphor of the iceberg: the portion apparent to the subject is just a small piece with respect to the whole structure. For this reason the *missing-ascription of ignorance* plays a role analogous to that of the epistemic bubble in the mechanism of creation and revision of beliefs. It assures a cognitive status of certainty about the agent ignorance that permits the agent to be confident in her choices and knowledge. The agent, not being able to see how much she ignores, considers the attainment of answers concerning her doubts a concrete way to remove her ignorance piece by piece.

The role of confidence is part of the autoimmune mechanism as much as the proper ignorance embubblement. The embubblement allows the agent to consider what is part of her doubts as the *entire amount of her ignorance* and her purpose will be to remove it as much as possible. In this sense, the role of the missing-ascription of ignorance is fully motivational. But the more effective consequence in the agent's cognition is the self-representation that the agent constructs in first-person perspective: indeed, there is a tendency to consider the knowing or ignorant self as a controllable part. The agent is fully aware of both the state of belief and of doubt, which are the only vehicles for her attainment of propositional/sentential knowledge and her partial awareness of ignorance. In the following subsection we will show that these partial recognitions drive the agent to formulate a sort of Homunculus Fallacy, when she tries to depict her epistemological state.

5.1 Cognitive Autoimmunity: The Homunculus Fallacy

In the case of the first-person perspective we have illustrated above the epistemic bubble provides two main illusions. The first illusion is strictly related to the epistemic dimension of the bubble: it provides the belief-based ascription of knowledge even when that knowledge is not entirely attained. The second illusion is related to the cognitive and emotional outcome of the bubble: it makes the agent convinced of being aware of the knowledge she possesses, even when she should not.

The same deceptive double effect also emerges from the missing-ascription of ignorance. On the one hand, it provides the agent the conviction that she is ignoring just a specific sort of data, categorizable in the framework of her first-person perspective. On the other hand it gives the agent the illusion of being able to have a clear view of her own ignorance. In both cases the agent is naively assured about her cognition. She thinks herself able to see her knowledge and her ignorance as they were, respectively, sets of attained or missing propositions. The agent is deluded into being, absurdly, in an *indifferent* position about her own ignorance/knowledge structure. This effect can be pictured as a sort of Homunculus Fallacy. The subject thinks herself almost as a double being: one part of her knows and ignores and another part can spot how much she knows and how much she ignores. Ironically, the constitution of the state of belief and of doubt can let the agent speak as if she possesses one information without actually having it, and viceversa. In order to make an example, if we think of a sentence like "If I knew how far is Paris from here, I could organize a trip for the week end", we are imagining to have an information that we do not possess. Viceversa, we can think something like "If I hadn't known that my wife was cheating on me I would still be with that harpy", where we can imagine to ignore something when we actually have that information.

The fairly hidden Homunculus Fallacy is clear: the autoimmune mechanism suggests that the agent can judge about the attainment of knowledge or the perception of ignorance, as if the judgment belonged to a distinct part, which directly knows or ignores. This illusory distinction allows us to also consider the property of *just apparent corrigibility* of the bubbles:

PROPOSITION 9 (Apparent corrigibility). Since each of us is in his own epistemic bubble, the distinction between merely apparent correction and genuinely successful correction exceeds the agent's command.

COROLLARY 9 (a) As previously stated, the cognitive agent from his own first-person perspective favors the option of a genuinely sound correction.

COROLLARY 9 (b) Within an epistemic bubble the distinction between belief-change and belief-correction is also "resolved" in favor of the latter (Woods 2005, p. 741).

When the agent realizes that the belief she had was incorrect, or the knowledge she thought she had was illusory, the change of mind does not break the mechanism of the bubbles. Since she has to replace an information with another one and the only way to do it is to believe she has gained a correct one, she simply *shifts* from a bubble to another, maintaining the autoimmune mechanism unbroken. The bubble was not corrected, it just changed. The homunculus fallacy helps this dynamic because, for the agent, the change of mind is seen as a correction of a wrong statement (a mere belief) with a truthful one (knowledge) as she was able to spot the difference from the first-person perspective. We can see, from a third-person perspective, that the transition is from a belief to another one but this perception is unaffordable by the self-assured agent.

As it can be imagined, a similar structure is present in the account of ignorance-based bubbles: the "end" of a missing-ascription of ignorance that happens when the obtained answer to a given doubt is just apparent. The missing-ascription of ignorance shifts to another problem, which arises in the presence of new collected information. While for the epistemic bubble there is a distinction between belief-change and belief-correction, which is resolved in favor of the latter, in the case of the ignorance-based bubble there is a distinction between change of doubt and ignorance-removal that is resolved in favor of the second. In conclusion, the autoimmune system provides the agent with an efficient and improvable mechanism of belief and doubt change without the loss of confidence in self-awareness.

6 Conclusion

The introduction of the problem of ignorance in the framework of a naturalization of logic involves problematic issues regarding the epistemological status of the "real agent". First of all we have added to Woods' Fallibilist principles what we have called the *negative affirmation*: this move rendered possible the examination of the naiveness of the individual agent about her own cognition, shifting the attention to the state of doubt (defined by Peircean dynamic) instead of belief. Thanks to this change of perspective, a new subtle reinterpretation of Woods' "epistemic bubble" has favored the elicitation of that *autoimmune mechanism* that affects not only the system of belief creation and revision of a human agent—considered not able to distinguish what she knows and what she only thinks she knows—but also the relationship between doubt and ignorance-recognition. As belief is "the condition of knowledge and the impediment of its attainment" (Woods 2005), doubt is the requirement that permits the emerging of uncertainness while preventing the integral cognition of the agent's ignorance.

Notably, we were able to reconsider Woods' "Enough Already Thesis" as one of the major effects of what we have called the Homunculus Fallacy, which affects both the ignorance and knowledge recognition of the agent. The fact that the Enough Already Thesis remains intuitively and practically effective is, in fact, strictly connected to the immunity that human cognition has from its own boundaries. We prosper and survive despite (or thanks to) our immunity from a fully aware state of our knowledge and ignorance. Hopefully further research concerning the immunized and ignorant part of human cognition will provide interesting new insights able to enhance the newborn field of the naturalization of logic, as much as the study of "errors of reasoning" has had so far.

References

Aliseda, A. (2005). The logic of abduction in the light of Peirce's pragmatism. *Semiotica, 1/4*(153), 363–374.

Dunbar, G. (2008). Towards a cognitive analysis of polysemy, ambiguity, and vagueness. *Cognitive Linguistics, 12*(1), 1–14.

Gabbay, D., & Woods, J. (2001). The new logic. *Logic Journal of the IGPL, 9*, 141–174.

Gabbay, D. M., & Woods, J. (2003). *Agenda relevance: A study in formal pragmatics. A practical logic of cognitive systems* (Vol. 1). Amsterdam: Elsevier.

Gabbay, D.M., & Woods, J. (2005). *The reach of abduction: Insight and trial. A practical logic of cognitive systems* (Vol. 1). Amsterdam: Elsevier.

Gigerenzer, G. (2004). Gigerenzer's law of indispensable ignorance. Published on *The Edge*, retrieved on https://edge.org/response-detail/10224.

Gigerenzer, G., & Goldstein, D. G. (1996). Reasoning the fast and frugal way: Models of bounded rationality. *Psychological Review, 103*(4), 650–669.

Gigerenzer, G., Todd, P., & The ABC Research Group. (1999). *Simple heuristics that make us smart*. Oxford: Oxford University Press.

Grice, H. P. (1975). *Logic and conversation*. Cambridge: Harvard University Press.

Hamblin, C. L. (1970). *Fallacies*. London: Methuen.

Hegel, G. W. F. (1977). *Phenomenology of spirit* [1807]. trans. by A. V. Miller, Oxford: Oxford University Press.

Ippoliti, E. (2015). Reasoning at the frontier of knowledge: Introductory essay. In E. Ippoliti (Ed.), *Heuristic reasoning*. Heidelberg: Springer.

Magnani, L. (2011). *Understanding violence. Morality, religion, and violence intertwined: A philosophical stance*. Berlin: Springer.

Magnani, L. (2013). Is abduction ignorance-preserving? Conventions, models, and fictions in science. *Logic Journal of the IGPL, 21*, 882–914.

Magnani, L. (2014). Are heuristics knowledge-enhancing? Abduction, models, and fictions in science. In E. Ippoliti (Ed.), *Heuristic reasoning* (pp. 29–56). Heidelberg: Springer.

Magnani, L. (2015). Naturalizing logic. Errors of reasoning vindicated: Logic reapproaches cognitive science. *Journal of Applied Logic, 13*, 13–36.

Magnani, L., & Bertolotti, T. (2011). Cognitive bubbles and firewalls: Epistemic immunizations in human reasoning. In L. Carlson, C. Hölscher & T. Shipley (Eds.), *CogSci 2011, XXXIII Annual Conference of the Cognitive Science Society*. Boston MA: Cognitive Science Society.

Moore, B. N., & Parker, R. (2012). *Critical thinkings*. New York: McGraw-Hill.

Peirce, C. S. (1931–1958). *Collected papers of Charles Sanders Peirce*. Cambridge: Harvard University Press. (Eds. by C. Hartshorne & P. Weiss, (Vols. 1–6). Ed. by A.W. Burks, (Vols. 7–8)).

Peirce, C. S. (1992–1998). *The essential Peirce. Selected philosophical writings*. Bloomington and Indianapolis: Indiana University Press. (Ed. by N. Houser & C. Kloesel (Vol. 1), Ed. by the Peirce Edition Project (Vol. 2))

Peirce, C. S. (1998a). How to make our ideas clear. In N. Houser & C. Kloesel (Eds.), *The essential Peirce. Selected philosophical writings* (Vol. 1). Indiana: Indiana University Press.

Peirce, C. S. (1998b). The fixation of belief. In N. Houser & C. Kloesel, (Eds.), *The essential Peirce. Selected philosophical writings* (Vol. 1). Indiana: Indiana University Press.

Polanyi, M. (1966). *The tacit dimension*. London: Routledge & Kegan Paul.

Proctor, R. N. (2005). Agnotology. A missing term to describe the cultural production of ignorance (and its study). In R. N. Proctor, (Ed.), *Ignorance* (pp. 1–36). Stanford: Stanford University Press.

Simon, H. A. (1993). Altruism and economics. *The American Economic Review*, *83*(2), 156–161.

Simon, H. A. (1997). *Models of bounded rationality*. Cambridge: MIT Press.

Walton, D. N. (1995). *A pragmatic theory of fallacy. Studies in rhetoric and communication*. Alabama: University of Alabama Press.

Woods, J. (2005). Epistemic bubbles. In S. Artemov, H. Barringer, A. Garcez, L. Lamb & J. Woods (Eds.), *We will show them: Essay in honour of Dov Gabbay* (Vol. II). London: College Publications.

Woods, J. (2007). The concept of fallacy is empty: A resource-bound approach to error. In L. Magnani & L. Ping (Eds.), *Model-Based Reasoning in Science, Technology, and Medicine*. Berlin: Springer.

Woods, J. (2013). *Errors of reasoning. Naturalizing the logic of inference*. London: College Publications.

Woods, J., Irvine, A., & Walton, D. (2004). *Argument: Critical thinking logic and the fallacies*. Toronto: Prentice Hall.

Towards a Caricature Model of Science

Woosuk Park

Abstract In Park (2014), I claimed that analogy between idealizations in science and caricatures in art might indicate a way toward a unified theory of representation. The basic idea for the analogy was secured by referring to Hopkins (1998) and Blumson (2009), who have discussed examples of pictorial misrepresentation by sharing caricatures and wanted-for posters of criminals. At least, this analogy may appease Hopkins's worry about how it is possible to see in a caricature of Blair both Blair and an enormous-mouthed thing, or how to see one set of marks as resembling both. For, just as idealizations in science can misrepresent and represent at the same time, caricature of Blair can achieve misrepresentation (enormous-mouthed thing) and representation (Blair) at the same time. Surprisingly, Niiniluoto (1997, 1999) used precisely the same examples in his account of reference by truthlike scientific theories (see Niiniluoto 2014). Encouraged by all this, I shall try to fathom what a caricature model of science would be like. I shall discuss what Niiniluoto means by "caricature model (or theory) of reference", and how this model (or theory) works within his theory of verisimilitude. In order to understand all this against a broader background, I shall also discuss both Goodman's analogy between descriptions in science and pictorial representations and Tomas Kulka's analogy between Popperian philosophy of science and quantitative model of aesthetic evaluation. Then, following the lead of Gombrich, I shall discuss some philosophical issues related to caricatures. Finally, by synthesizing all these discussions, I shall present the analogy between idealization and caricature in more detailed fashion.

1 Introduction

In Park (2014), I claimed that analogy between idealizations in science and caricatures in art might indicate a way toward a unified theory of representation. The basic idea for the analogy was secured by referring to Hopkins (1998) and Blumson

W. Park (✉)
Humanities and Social Science, Korea Advanced Institute of Science and Technology (KAIST), Daejeon, Korea
e-mail: woosukpark@kaist.ac.kr

L. Magnani et al. (eds.), *Philosophy and Cognitive Science II*,
Studies in Applied Philosophy, Epistemology and Rational Ethics 20,
DOI 10.1007/978-3-319-18479-1_5

(2009), who have discussed examples of pictorial misrepresentation by examples of caricatures and wanted-for posters of criminals. At least, this analogy may appease Hopkins's worry about how it is possible to see in a caricature of Blair both Blair and an enormous-mouthed thing, or how to see one set of marks as resembling both. For, just as idealizations in science can misrepresent and represent at the same time, caricature of Blair can achieve misrepresentation (enormous-mouthed thing) and representation (Blair) at the same time. Surprisingly, Niiniluoto (1997, 1999) used precisely the same examples in his account of reference by truthlike scientific theories (see Niiniluoto 2014). Encouraged by all this, I shall try to fathom what a caricature model of science would be like.[1]

In Sect. 2, I shall briefly explain how naturally the analogy between idealizations in science and caricatures in art presents itself. In Sect. 3, I shall discuss what Niiniluoto means by "caricature model (or theory) of reference", and how this model (or theory) works within his theory of verisimilitude. It will become evident that, in order to flesh out the idea of caricature model of science, we not only need to locate it against the background of comparing science and art, but also have to learn more about caricatures. For the former, Sect. 4 will be devoted to compare the analogy between idealizations in science and caricatures in art with both Goodman's analogy between descriptions in science and pictorial representations and Tomas Kulka's analogy between Popperian philosophy of science and quantitative model of aesthetic evaluation. For the latter, in Sect. 5, I shall discuss the role and function of Gombrich's treatment of caricatures in his *Art and Illusion*. In Sect. 6, I shall hint at how to deepen our understanding of the analogy between idealization in science and caricature in art.

2 Misrepresentation in Context

Godfrey-Smith presents the problem of misrepresentation as the problem for naturalistic theories for mental representation (Godfrey-Smith 1989). However, he didn't even mention the resemblance theories as one of the naturalistic theories of mental representation. Fodor does mention the resemblance theory. But he excludes the resemblance theory as definitely wrong at the outset. So, we seem to have a very good reason to believe that the problem of misrepresentation is peculiar to the informational theories of mental representation. In other words, according to the assumption presumably widespread in the discussions of mental representation, the problem of misrepresentation does not arise for the resemblance theory. On the other hand, both Hopkins and Blumson discuss the problem of misrepresentation for the resemblance theory in aesthetics. Further, they seem to presuppose that this problem is peculiar to the resemblance theory. What is going on?[2]

[1] As one anonymous reviewer points out, this paper is mainly exploratory and programmatic, and therefore unavoidably of somewhat speculative character.

[2] This section is based on Park (2014).

My hunch is that Goodman's celebrated rejection of resemblance theory is largely to blame for these situations in philosophy of mind and aesthetics. According to Goodman, unlike representation, resemblance is reflexive and symmetric. Without any further argument other than this seemingly undisputable observation, Goodman claims that resemblance is neither necessary nor sufficient for representation in art. Examples Goodman exploits in favor of the insufficiency of resemblance for representation are the automobiles off an assembly line and the twin brothers. In these cases, "neither one of a pair of very like objects represents the other". (Goodman 1976, p. 4). In order to prove that resemblance is not necessary for representation, Goodman simply resorts to the widely accepted fact that "almost anything may stand for almost anything else". [Ibid., p. 5]. Insofar as we can accept that "denotation is core of representation", Goodman seems to have an extremely strong, if not conclusive case for his conclusion that representation is "independent of resemblance" [Ibid.]. The power of Goodman's terse argument for the independence of representation from resemblance turns out to be formidable. To testify its widespread and untouchable influence, let it suffice to give a few examples: Lopes (1996), Bailer-Jones (2009), Fodor (1984), Park (2014). Without further ado, we may safely count Goodman's treatment of representation as "the twentieth century's seminal text on the matter". (Van Fraassen 2008, p. 16).

What is remarkable is that the situation seems rather different in philosophy of science. The problem of verity still persists in philosophy of science, at least with the problem of verisimilitude. What we need to understand is the reason why the problem of misrepresentation apparently does not arise in philosophy of science, especially in connection with the huge literature concerning verisimilitude. In this regard, it seems useful to turn to Chakravartty's recent discussion of analogies between science and art. Chakravartty (2010) seems to be a rare case for appropriating the inspirations from art to the problem of truth and representation in science. In order to understand "how and in what manner scientific representations can be true", he believes, analogies to representation in art "can serve as valuable heuristics". [Ibid., p. 33]. But how could that be the case? Chakravartty assumes that, in order to invoke truth in science, one should be able to make sense of "the idea that inaccurate representations can be closer to the truth". [ibid., p. 34]. But, in order to secure a sound notion of approximate truth, he believes, one needs to elaborate "conditions of approximation". And his basic idea is using the analogy between science and art for understanding abstraction and idealization in science in order to illuminate the conditions of approximation [ibid., p. 35].

Interestingly, Chakravartty depends crucially on some features of Goodman's conventionalism on representation in his employment of the analogy between art and science. For he writes:

> The contrast between depiction and mere denotation as a central feature of representation in art is an analogy for the contrast between truth and mere reference as a central feature of representation in the sciences. Higher degrees of approximate truth can be understood in terms of improved representations of the natures of target systems in the world, and this improvement can be mapped along two dimensions: how many of the relevant properties and

> relations one describes (abstraction), and how accurately one describes them (idealization) [ibid. p. 45].

Here, from Chakravartty's introduction of abstraction and idealization, we seem to have another evidence for the observation that, unlike in philosophy of mind or aesthetics, existence of misrepresentation is duly appreciated in philosophy of science. Indeed, there is huge literature in philosophy of science that discusses the problems of idealization and abstraction in individual sciences. Martin R. Jones, for example, explicitly discusses abstraction and idealization in connection with misrepresentation:

> On the regimentation of usage I am thus proposing, the term 'idealization' applies, first and foremost, to specific respects in which a given representation misrepresents, whereas the term 'abstraction' applies to mere omissions (Jones 2005, p. 174).

In masterly fashion, Jones explains the difference between idealization and abstraction by using the example of the flight of the cannonball. According to the usual model for the flight of the cannonball, for example, "the gravitational force due to the Earth has the same magnitude and direction at all points on the cannonball's directory, whereas in fact both the magnitude and the direction of that force will vary". Jones calls some such misrepresentation an "idealization". [ibid., p. 182]. On the other hand, the model simply omits certain features of the system "without thereby misrepresenting or distorting the system". [ibid., p. 184]. For example, it does not mention the composition, its internal structure, its color, or its temperature. These are cases of what Jones means by "abstraction".

3 Niiniluoto's Treatment of the Caricature Theory of Reference

Niiniluoto nicely introduces the problem of reference fixing for theoretical terms in science within the larger context of the debate between scientific realism and instrumentalism. And, here he suggests a "caricature model (or theory) of reference". As is well-known, descriptive theory and causal theory are the two major rival theories of reference for proper names. When applied to theoretical terms, Niiniluoto finds the causal theory problematic. He finds it hopeless to originally give a name in the manner of causal theory "to an unobservable elementary particle or mental event. Further, causal theory fails to account for reference failure by allowing too much reference invariance. Niiniluoto cites 'phlogiston', 'ether', and 'demon' as examples for this. So, Niiniluoto agrees with Nola (1980) and Kroon (1988) that "reference fixing needs some descriptive elements". And he understands this as leading us "toward an account of theoretical terms as 'implicitly defined' or cluster concepts' (Putnam 1975)". [Ibid.].

Starting with a Fregean descriptive account for general terms (DR1), Niiniluoto successively presents more improved formulations, (DR2) through (DR6). However,

I shall focus on how he moves from (DR2) to (DR3) and (DR4). For, it is his discussion of (DR2), where he introduces "caricature model of science", and he finds the superiority of (DR3) and (DR4) over (DR2) in that they are more precise formulations employing "the concepts of approximate truth AT and truthlikeness Tr (Niiniluoto 1987a)":

> (DR1) A term t occurring in theory T refers to object b iff b satisfies the claims of T containing t.
> (DR2) A term t occurring in theory T refers to object b iff b satisfies the majority of the claims of T containing t.
> (DR3) Term t occurring in theory T refers to the actual object b which maximizes AT(T,b)
> (DR4) Term t occurring in theory T refers to the actual object b which maximizes Tr(T, b).
> [Ibid., pp. 126–129]

Niiniluoto, as a critical scientific realist, cannot accept descriptive theory. For (DR1) "would have catastrophic consequences for scientific realism": i.e., (1) excluding the possibility of reference invariance, and (2) reference failure of most scientific theories [Ibid., p. 126]. According to Niiniluoto, (DR2) results by rejecting (DR1) but preserving its correct core. However, he is skeptical about its ability to cover "typical cases where a theory is false but 'close to the truth' [Ibid., p. 127].

At this crucial stage, Niiniluoto introduces Putnam (1975) Principle of Charity (or Principle of benefit of Doubt), which "allows that the person proposing a term may make reasonable changes in the original description". Given this principle, as Niiniluoto claims, "Bohr referred precisely to those entities by his term 'electron' [Ibid., p. 128]. Further, Niiniluoto finds "another form of this idea" in Lewis (1970):

> if a theory T is not realized by any entity, it may still have a 'near-realization' which realizes another theory T' obtained from the original theory T 'by a slight weakening or a slight correction' (p. 432). Then the theoretical term in T denotes the entity which is 'the nearest near-realization' of the theory, if it 'comes near enough' [Ibid., p. 128].

Here, Niiniluoto finally introduces caricatures to his discussion:

> When applied to singular reference, this treatment could be called the caricature theory of reference. A good caricature of a person b distorts the outlook of b by an amusing exaggeration of the features of b, but still it must bear sufficient similarity to its model so that we easily see the reference to b (Niiniluoto 1999, p. 128).

It is not clear whether Niiniluoto here has some special reasons for confining caricature model only to singular reference. At least in spirit, however, he may allow extending the caricature model to general or theoretical terms in science. For, he recently reaffirms the possibility of comparing scientific idealization to caricature:

> As caricatures to some extent misrepresent their targets, their ability to refer to their targets is denied by Fregean descriptive theories of reference, which require that a theory can refer only to those entities which it correctly describes. However, if we adopt a principle of charity to the effect that a theory refers to those objects which it describes in the most truthlike manner, then such caricatures can refer to their targets (see Niiniluoto 1997). The possibility of reference failure is explained by choosing a threshold or a lower value for the required degree of truthlikeness (Niiniluoto 2014, p. 378).

The obvious merit of Niiniluoto's approach lies in the fact that it promises us to allow a quantitative, thereby objective, assessment of rival hypotheses or theories. If it is possible to make an analogical case for caricatures, one natural consequence would be that we may rank all the different caricatures of the same object in terms of the degree of truthlikeness. A closely related point is that we may extract from such a possibility a potentially damaging criticism of Goodman-Elgin type treatment of caricatures as merely instances of "representation-as". (Elgin 2009). I will return to this topic below.

Another interesting thing to note is that here Niiniluoto seems to assume that there is consensus regarding what caricatures are. For, he does not discuss this problem at all. I am not accusing him for any obvious mistakes in his understanding of caricatures. My point is rather that he implicitly indicates the possibility of improving our understanding of the problem of reference fixing from our relatively unproblematic views about caricatures. If this is not his intention, what is it for to dub Putnam-Lewis approach to reference as "the caricature model of reference"? Niiniluoto does not discuss explicitly how he understands caricatures in art. As a consequence, it is not clear what we can learn from his dubbing of Lewisian type of treating theoretical terms in science as "caricature model (or theory) of reference". Be that as it may, it may not be trivial to note that Chakravartty and Park (2014) seem to reveal a similar optimism for securing lessons from caricatures for understanding science. Park (2014) suggests a way of solving Hoffman's puzzle about caricatures by learning something from abstractions and idealizations in science. But Niiniluoto, Charkravartty, and Park (2014) also want to learn more about abstractions and idealizations in science by appealing to caricatures. Aren't they falling in a circle in this regard?

Niiniluoto also presents an extremely perceptive observation on the caricature model of reference:

> In fact, the caricature model of reference as such is not quite appropriate for science, since theories are not pictures of already known objects (such as the drawings of Charles de Gaulle and Elvis Presley). Rather they are attempts to describe some so far unknown theoretical entities in the basis of incomplete, partial, uncertain, and indirect information. In this sense, they can be compared to the pictures of unknown murderers sometimes published by the police, drawn by an artist or by a computer relying on the evidence provided by eyewitnesses. (The picture of the still unidentified assassin of prime Minister Olaf Palme is a case in point) (Niiniluoto 1999, p. 132).

In view of the scientific realists' focal interest in the reference of theoretical terms, it cannot be a small point to realize that the case of unknown theoretical entities in science is more similar to the case of the pictures of unknown criminals than the case of caricatures of well-known individuals. Park (2014) also cites the example of the wanted-for posters of criminals referring to Blumson (2009). However, it is merely for stimulating the readers' interest in the puzzles involved. There is no positive contribution to our understanding of the puzzles other than complaints of a frustrated mind. Therefore, though brief, Niiniluoto's suggestive remarks in the quote above deserves careful attention.

Nonetheless, we not only need to locate the lessons from Niiniluoto's caricature model of reference against the broader background of comparing science and art, but also have to learn more about caricatures. The next two sections will serve for these concerns.

4 Too Many Analogies Between Art and Science?

In order to appreciate the true value of the analogy between idealizations in science and caricatures in art, it seems advisable to compare it with other analogies between science and art. In fact, there already seem to be quite a lot of some such analogies suggested to date, including Chakravartty's. For my present purpose, however, it would be enough to examine just two of them: i.e., Goodman's analogy between scientific description and pictorial representation, and Kulka's analogy between Popperian philosophy of science and quantitative model of aesthetic evaluation.[3]

Goodman's influence on aesthetics and the philosophy of art in the last four decades has been so salient that it even struck a historian-critic "as standing in relation to classical aesthetics as relativity theory does to classical physics". (Ackerman 1981, p. 249). In some sense, however, the alleged success of Goodman is only partial, for we fail to appreciate what exactly is so revolutionary in Goodman's understanding of art. We have never been quite serious about uncovering the full implications of his rejection of the conventional understanding of art and science (Goodman 1972, p. 83). Nor do we pay due attention to Goodman's thesis that the primary function of art is cognitive, even if we can witness the recent surge of interest in cognitive functions of art (Goodman 1976, p. 258).

The primary task of Goodman in the first chapter of *Languages of Art* is soley to pave the way toward the analogy of science and art by emphasizing the analogy of description and representation. Ironically enough, he is so successful in his emphasis on the analogy that he cannot differentiate description and representation until he introduces the general theory of symbols in Chap. 4. Only after having presented such a theory, does he tackle the problem of comparing art with science. Consequently, he arrives at the conclusion that aesthetic experience is, like science, a form of understanding, and that the difference between art and science is not the difference between the emotive and the cognitive, but rather a difference in domination of certain specific characteristics of symbols (Goodman 1976, p. 264).

The point of departure for the analogy between pictorial representation and verbal description is provided by the nominalistic idea which regards "denotation" as "the core (or the necessary condition) of representation". (Goodman 1976, p. 5). According to Goodman, in order to represent an object, "a picture must be a symbol

[3]As is pointed out by one anonymous reviewer, this paper fails to define one of its central concepts, i.e., the concept of analogy. This could be a serious omission, indeed. However, it must be beyond the scope of this paper to clear up the on-going controversies around the concept of analogies. Interested readers may find Shelley (2003) as a nice point of departure.

for it, stand for it, refer to it". (Goodman 1976, p. 5). The popular view that appeals to resemblance is untenable because the relation of resemblance is, unlike that of representation, reflexive and symmetric (Goodman 1976, p. 4). Thus, resemblance is neither a necessary nor sufficient condition for representation.

The initial difficulty for such a nominalistic theory of meaning which views denotation as a necessary condition emerges from the paradoxical case of "null-denotation". "Since there is no Pickwick and no unicorn, what a picture of Picwick and a picture of a unicorn represent is the same. Yet surely to be a picture of Pickwick and to be a picture of a unicorn are not at all the same". (Goodman 1976, p. 21).

Goodman solves this problem by mobilizing his old distinction between primary extension and secondary extension. Goodman calls "the extension of a predicate by itself its *primary* extension, and the extension of any of its compounds a *secondary* extension". Thus, the criterion of the likeness of meaning is formulated as follows: "two terms have the same meaning if and only if they have the same *primary* and *secondary* extensions". (Goodman 1972, p. 227). Then, now we can understand the difference between saying what the picture denotes and saying what kind of picture it is (Goodman 1976, p. 22). "A picture must denote a man to represent him, but need not denote anything to be a man-representation" (Goodman 1976, p. 25).

Consequently, the problem of null-denotation leads us to the insight about the classificatory function of representation. "Just as objects are classified by means of, or under, various verbal labels so also are objects classified by or under various pictorial labels". (Goodman 1976, pp. 30–31). Further, since representation is a matter of classifying and characterizing objects, we can understand that it is not a matter of passive reporting, but a matter of organizing and remaking our world (Goodman 1976, pp. 31–33)."Effective representation and description require invention". (Goodman 1976, p. 33).

Goodman thinks that the characterization of representation as denotation dependent upon pictorial properties is too ad hoc to be accepted as final (Goodman 1976, p. 225). For this reason, Goodman can treat the problem of distinguishing representation from other modes of denotation only after he analyzes the symbol systems. According to him, a notational system satisfies five requirements (2 syntactic requirements and 3 semantic requirements), and a language satisfies at least the first two syntactic requirements. Thus, "nonlinguistic systems differ from language … primarily through lack of differentiation—indeed through density (and consequent total absence of articulation)—in the symbol systems". And he goes on to say that "a scheme is representational only insofar as it is dense; and a symbol is a representation only if it belongs to a scheme dense throughout or to a dense part of a partially dense scheme". (Goodman 1976, p. 226).

Apparently, there is one salient similarity between Chakravartty's and Goodman's analogies between science and art. Above all, they are analogies between representation in science and art. The crucial difference between their analogies is also apparent. Unlike that of Chakravartty's, which appeals to truthlikeness, Goodman's analogy is suggested with the complete exclusion of resemblance between what represents and what is represented. This might be the reason why Goodman invokes the analogy between description and pictorial representation rather than the analogy between scientific and pictorial representation. For, in such a formulation, the conventional character of linguistic representation could be highlighted. Be that as it may, it seems pertinent to fathom what consequences would follow such a difference. In the case of Goodman and Elgin, there seems to be no room for comparing idealizations in science and caricatures in art. For, there is no quantitative (or objective) standard according to which once can differentiate different degrees of truthlikeness (or similarity). On the other hand, as we saw above, it is Chakravartty's focal interest to figure out the criteria according to which one can compare different idealized representations in science and caricatures in art.

The fundamental insight of Kulka is that we may meaningfully discuss "testability or justifiability of aesthetic value judgements". By appealing to an intuitive notion of "alteration", Kulka suggests as follows:

> Alternatives which are inferior to a given work of art could be seen as confirming evidence for claims about its rightness, while superior alterations would count as disconfirming instances of such aesthetic value judgements (Kulka 1989, p. 202).

Further, he claims that "comparative judgements of alterations are considerably more simple than judgements of the aesthetic value of works of art". [Ibid.]. Whether or not Kulka's claim is backed up by scientific statistical data, his idea is that such aesthetic judgements alterations can be called "basic aesthetic judgements" that can play the role analogous to the role of "basic statements" in Popperian philosophy of science.

It is hard to understand why Kulka does not discuss Gombrich's views in any connection with Popperian aesthetics. For, not to mention their lifelong friendship, Gombrich himself explicitly pays tribute to Popper's strong influence on *Art and Illusion*. As is well-known, Gombrich's theory of schema and correction seems to be following Popper's hypothetico-deductivisim. Gombrich also indicates direct connection of his views with Popper's searchlight theory or method of trial and error. There is no doubt that Kulka's strategy of finding analogy between Popperian falsificationist philosophy of science and aesthetic evaluation in art is quite original and invaluable in itself. In a sense, Kulka's analogy is ad hoc. For, it seems to be motivated to avoid endorsing Gombrich's Popperian aesthetics. Without invoking the problem of evaluation in art, it is perfectly possible to develop Popperian aesthetics or philosophy of art. On a par with Chakravartty's and Goodman's analogies between art and science, Kulka could have presented Popperian analogy between science and art, by focusing on representation in science and art. If such a choice had been made by Kulka, it would have been extremely difficult, if not impossible, not to discuss Gombrich's Popperian aesthetics. It seems even possible that Kulka might have arrived at a position exactly like that of Gombrich's.

5 Gombrich as Philosopher: The Role of Caricatures in Art and Illusion

Be that as it may, we have another very weighty reason to look into Gombrich's views on the analogy between science and art. For, in *Art and Illusion*, Gombrich devotes an entire chapter to the problems of caricatures. What exactly are the roles and functions of this chapter entitled "Chapter X. The Experiment of Caricature"?

Together with Ernst Kris, who was the supervisor of his doctoral dissertation, Gombrich did extensive research into the history of caricature for a long period of time. Since caricature is one of Gombrich's life-long concerns, scholars and historians may count this chapter as an indispensable piece for solving the puzzle of his intellectual development. However, uninitiated readers might simply be curious about why comic pictures should be discussed so extensively in a volume on the history of pictorial representation. In fact, we find the clues for both puzzles in *Art and Illusion* itself:

> Our starting point at the time was the question of why portrait caricature, the playful distortion of a victim's face, makes only so late an appearance in Western art. The word and the institution of caricature date only from the last years of the sixteenth century, and the inventors of the art were not the pictorial propagandists who existed in one form or another for centuries before but those most sophisticated and refined of artists, the brothers Carracci [Ibid., p. 343].

Scholars and historians who are interested in Gombrich's intellectual development may find Popper's influence in the same chapter very useful. For example, Gombrich's discussion of Freudian psychoanalysis must be instrumental for understanding exactly where he distanced himself from his former teacher and collaborator. Also, Gombrich's discussion of some of the most ambitious programs in the history of art, such as Rembrandt's program, would be powerful enough to amend any vulgar prejudice underestimating caricature.

Now let me select from Gombrich's thought-provoking discussions some passages that might have implications for my current issue, i.e., the analogy of idealization and caricature. The point of departure for Gombrich to answer his question as to the late arrival of the genre of portrait caricature seems to be the insight embedded in the following:

> Things objectively unlike can strike us as very similar, and things objectively rather similar can strike us as hopelessly unlike [Ibid., p. 331].

Also, Gombrich's ultimate answer can be found in the following remark:

> The invention of portrait caricature presupposes the theoretical discovery of the difference between likeness and equivalence [Ibid., p. 343].

Further, we can gather more information about the difference between likeness and equivalence in the following passage:

> IN THIS FORMULATION caricature becomes only a special case of what I have attempted to describe as the artist's test of success. All artistic discoveries are discoveries not of likenesses but of equivalences which enable us to see reality in terms of an image and an image in terms

> of reality. And this equivalence never rests on the likeness of elements so much as on the identity of responses to certain relationships [Ibid., p. 345].

If I am on the right track, then the next obvious question for Gombrich would be "exactly when, by whom, and how the discovery of the difference between likeness and equivalence was made". Even though it is simply impossible for me to report how Gombrich would answer this question, the following quote may hint at a broad outline of what it would be like:

> In our story, Hogarth stands somewhere in between Leonardo and Le Brun on the one hand—both of whom he quoted—and Töpffer on the other. To Leonardo, nature was still the great teacher and rival and the training of memory was just a by-product of his interest in morphology. For le Brun, art had become a lofty language from which it was dangerous to depart without loss of caste. Hogarth accepted the idea of art as a language and seized eagerly on the possibilities it offered for the creation of characters with which to people his imaginary stage.
>
> That this was his aim is apparent from such prints as *Characters and Caricaturas* [288], which drives home the difference between mastery of variety—the knowledge of character—and the exaggerations of caricature. Later in his life he defined this difference explicitly. Caricature rests on comic comparison. ... Character, by contrast, rests on knowledge of the human frame and heart. It shows the artist as a creator of convincing types [Ibid., p. 350].

Gombrich rather extensively discusses Töpffer as the inventor and propagator of "picture story, the comic strip" [Ibid., p. 336], and as the author of *Essay du physiognomie* 1845 [Ibid., p. 340]. What is central in this discussion must be Töpffer's "psychological discovery that you can evolve a pictorial language without any reference to nature, without learning to draw from a model". [Ibid., p. 339]. If "a knowledge of physiognomics and human expression" is the only thing needed for the pictorial narrator in the lead of Töpffer, as Gombrich points out, Le Brun's patternbook and Hogarth's *Characters and Caricaturas* seem to have utmost importance in the history portrait caricature. It is especially so for my purpose of drawing an analogy between idealization and caricature.

According to Gombrich,

> Le Brun compiled a patternbook of typical heads [286] in the grand manner—the fierce soldier, the simpering maiden—and then proceeded to analyze these heads in order to find out what it was that made them expressive [Ibid., p. 348].

In order to appreciate the importance of patternbooks for Gombrich, we need to remember that, earlier in *Art and Illusion*, referring to popular books of our time to teach "how to draw trees" or "how to draw birds", he claims that "all these books work on the principle we would expect from the formula "schema and correction"". [Ibid., p. 147]. Also, we need to remember, such a discussion is led to "the philosophic distinction between "universals" and "particulars"". [Ibid., p. 152]. Hogarth's distinction between character and caricaturas, or Töpffer's distinction between "permanent traits" indicating character and the "impermanent ones" indicating emotion [Ibid., p. 340] can get further significance against Gombrich's discussion of schema and correction or of universals and particulars.

6 Idealizations and Caricatures

We are now in a much better position to present the analogy between scientific idealization and pictorial representation. Before plunging into the project of presenting that analogy in in-depth fashion, let me briefly introduce the problem of "caricature generator" first. The idea of caricature generator was first given in an MIT master thesis in computer science (See Brennan 1985). For the past three decades, we have seen its continued progress up to 3-D version of it. What is pertinent for my present purpose is that caricature generators promise to compare the resulting caricatures quantitatively:

> The program finds corresponding points on the face and the norm, represents the difference between these pairs of points on the face as a vector, and increases the length of the vector by a specified constant proportion, thereby shifting the face-point further from its corresponding norm-point (Rhodes 1996, p. 41).[4]

Given some sample product of caricature generators (see Fig. 1), no reader of *Art and Illusion* would fail to notice the possible relevance of Gombrich's discussion of *Le Poires* (see Fig. 2):

> But the *locus classicus* for a demonstration of this discovery of like in unlike is the *Poire* [282], the pear into which Daumier's employer, Philipon, transformed the head of the Roy Bourgeois, Louis Philippe. Poire means a "fathead," and when Philipon's satirical papers continuously pillories the King as a *poire*, the editor was finally summoned and a heavy fine was imposed (Gombrich 1960, p. 344).

There are certainly some obvious similarities between the caricature drawn by a caricature generator and *Le Poires*. The most salient similarity seems to be that in both we are supposed to have a sort of algorithm according to which we can systematically produce caricatures in varying degrees of resemblance with the original object. On the other hand, there seem to be also some differences between them. In the caricature generator's case, it is not clear which particular aspect is distorted or exaggerated. But we may point out a few aspects of *Le Poires* that are distinctively distorted or exaggerated: e.g., wrinkles, hair, and cheeks.

Now, we may rephrase the similarity between the caricature drawn by caricature generators and *Le Poires* in the language of idealization. Philosophers have called the reverse operation of idealization as concretization, de-idealization, or factualization. Niiniluoto seems to highlight the strength of his theory of truthlikeness in that it allows us to compare different hypotheses or theories by this degree of idealization and concretization (Niiniluoto 1999). In other words, we can compare two idealizations in science or two caricatures in art in terms of the degree of similarity. In this regard, we may simply encourage the interaction between science and art to learn more from the analogy between idealizations in science and caricatures in art.

---

[4]The major focus of Rhodes 1996 is, as its title "Superportraits: Caricatures and Recognition" indicates, the question of "whether caricatures can be superportraits". In the same place, Rhodes quotes Annibale Carracci' sremark as follows: "A good caricature, like every work of art, is more true to life than reality itself". (Rhodes 1996, p. 18).

On the other hand, the difference between the caricature drawn by caricature generators and *Le Poires* seems to indicate some of the toughest challenges. For, we are unable to identify which aspects are to be distorted or exaggerated in order to secure good or successful caricatures. Nor do we know what techniques of idealization or caricaturing are available. What becomes clear is that in both cases of idealization and caricature we are lacking any well-established classification. But, no meaningful interaction between science and art in connection with idealization and caricature can get off the ground without such classification.

Probably, recent discussion of idealization can provide us with at least a rough classification. McMullin's famous paper "Galilean Idealization" (McMullin 1985), where he distinguishes between (1) mathematical, (2) construct, (3) formal, (4) material, (5) causal, and (6) subjunctive could be a nice example. McMullin claims, however, that there are only two main forms of idealization:

> In construct idealization, the models on which theoretical understanding is built are deliberately fashioned so as to leave aside part of the complexity of the concrete order. In causal idealization the physical world itself is consciously simplified (McMullin 1985, p. 273; see also 255).

The so-called "construct idealization" involves "a simplification of the conceptual representation of an object" (Morrison 2005, p. 152). According to Morrison, it is "a more specific type of mathematical idealization", and "used in the development of models as opposed to the formulation of laws". As Morrison points out, construct idealization seems similar to what Chakravartty, Jones, and Cartwright call abstraction [Ibid.].

Further, according to McMullin, formal and material idealization are "two different aspects of a single technique", i.e., construct idealization (McMullin, p. 259). He distinguishes between them depending on whether what is simplified or omitted is "features that are known" or "features that are unspecified and deemed irrelevant to the inquiry at hand". (McMullin, p. 258). McMullin also divides causal idealization into two experimental and subjunctive idealization depending on whether "artificial ('experimental') context is constructed" or "performed in thought". (McMullin, p. 273).

As is evident from the subsequent discussions, e.g., Morrison (2005), McMullin's classification idealization seems to leave many deep problems untouched. For example, Morrison points out that "idealization in modern physics often involves constructing models that may have little to do with reality". (Morrison 2005, p. 151). According to Morrison, most accounts of idealization are also problematic in assuming that concretization "results in the model gradually becoming a more realistic representation of the phenomena". [Ibid., p. 153], for, this assumption oversimplifies the process of model building. Morrison claims that "physical concepts like that of the Higgs field are not only highly idealized but their degree of departure from real physical systems often cannot be determined *due to a lack of information*". (Morrison, p. 152; emphasis is mine).

The last point has utmost importance for Morrison, because it allows her to conclude that there are two kinds of idealization in physics:

> The first, which I will call "computational idealization," involves the straightforward sort of approximation used in cases like the rigid rod or frictionless plane, where we know how to account for perturbations and can calculate the degree of departure between the real and ideal cases. The second kind, predictive idealization, typically involves a variety of different ways of idealizing the phenomena, each of which is used for different purposes, so that as a result we are unable to determine the degree of approximation between the real system and the idealized model (Morrison, p. 158).

Now the question is whether we can at least find similar patterns in caricatures paralleling these different types of idealizations in science. As we saw above, we do not yet have a systematic classification of idealization. In view of the more than two-thousand-year history of science, we may well raise the question as to why the method of Galilean idealization appeared so late, as Gombrich and Kris were curious about the relatively late arrival of caricature in the history of art. However, the classification of idealization proposed by McMullin or Morrison can and should be the barometer for checking the situation in caricature. Can we strengthen the analogy between scientific idealization and pictorial representation in art in terms of the different techniques of idealization and caricaturing?

My pessimism about this possibility must already be apparent, for, I do not have any precedent in attempting a classification of caricature in terms of the techniques employed. Of course, we cannot exclude the possibility that some such attempts were made by one of the pioneers or historians of caricature. Even so, it is not well-known to the general public or to the philosophers of science. Nevertheless, the review above seems to indicate somewhat intriguing points of comparison. Morrison's classification of idealization into computational and predictive idealization may find counterparts in caricature. Insofar as we can systematically compare different products of caricature generation, they could be called "computational caricature". On the other hand, when we do not have any standard according to which we can base quantitative assessment except for relying on the beholder's subjective evaluations, we might be dealing with cases of "predictive caricature".

Also, some observations from caricatures might throw light on what seems unsatisfactory in the classification of idealizations. As we saw above, some scholars sharply distinguish between abstraction and idealization in science, but there are also others, who lump them together as idealization. That means, we need to rethink the alleged differences between abstraction and idealization. As it turns out, it seems extremely important to decide which aspects of the original are to be ignored or exaggerated. Now, insofar as both simple omission and exaggeration are distortions, we need to confirm whether there is no possible case where the former could be more radical distortion than the latter. If this idea makes sense, then we may have reason to rethink whether some simple diagrams in Euclidean geometry are not only abstractions, but also idealizations.

As we saw above, physionomics and facial expressions are utterly important in the development of caricature, and, we also know that there is a long tradition in art that counts the average face as the most beautiful.[5] If so, all the experiments reported by Gombrich leading toward the invention of caricature as a genre can be potential sources of insights for scientific idealization.

7 Concluding Remarks

One can witness the recent surge of interest in fictions in science (see Suárez 2009). This invites also a strong attack on fictionalism in science (see Magnani 2012; Woods 2014; Portides 2014; Giere 2009; Teller 2009). My sympathy with the latter should be obvious. I feel that there is severe danger involved in spreading fictionalism in science to the general public. However, I am a bit worried about the possibility that by rejecting factionalism in science we might underestimate the cognitive function of art. In other words, my endeavor in this paper might be counted as an effort to find a way out from this dilemma: If we grasp the first horn of factionalism in science, we might fall into a radical relativism; On the other hand, if we grasp the second horn of abandoning it once and for all, we might unduly ignore the cognitive function of art.

According to Suárez, we can distinguish between two different kinds of falsehood, thereby between *fictional* and *fictive* assumptions. The former refers to a case where "an assumption about some entity x is false, because there is no such entity in reality", while the latter refers to a case where "the same assumption is false, because the entity x is incorrectly described". (Suárez 2009, p. 13). Further, based on this distinction, he contrasts "a thorough or wide fictionalism" and "narrow fictionalism". According to the former, "fictive assumptions and representations *are* fictions". On the other hand, according to the latter, only the fictional assumptions and representations are fictions. [Ibid.]. The implication for this rivalry for the instrumentalist/scientific realist controversy is rather obvious, for, what is at stake is whether abstraction and idealization are fictions.

Against such a broader background, our discussion of the analogy between idealizations in science and caricatures in art might unexpectedly shed new light. Insofar as caricatures have a place in the history of pictorial representation, and there is a nice analogy between idealizations in science and caricatures in art, we may block any attempt to treat idealizations as fictions successfully.

Acknowledgments I am enormously indebted to anonymous referees' useful suggestions and criticisms of various sorts. As usual, Lorenzo Magnani and Ping Li provided me with the necessary moral support.

[5]Rhodes 1996 claims that "implicit in the notion of caricature as the exaggeration of distinctive information is the concept of a norm or reference point against which the exaggeration occurs". Further, she notes that "the best kind of norm for creating a recognizable portrait caricature is probably an average face that captures the central tendency of the population (or some relevant subset, such as young, female faces)". (Rhodes 1996, p. 19).

References

Ackerman, J. S. (1981). Worldmaking and practical criticism. *The Journal of Aesthetics and Art Criticism*, *39*, 249–254.

Bailer-Jones, D. (2009). *Scientific models in philosophy of science*. Pittsburgh: University of Pittsburgh Press.

Blumson, B. (2009). Images, intentionality and inexistence. *Philosophy and Phenomenological Research*, *79*(3), 522–538.

Brennan, S. E. (1985). Caricature generator: The dynamic exaggeration of faces by computer. *Leonardo*, *18*(3), 170–178.

Chakravartty, A. (2010). Truth and representation in science: Two inspirations from art. In R. Frigg & M. C. Hunter (Eds.), *Beyond mimesis and convention* (pp. 33–50). Boston: Springer.

Elgin, C. Z. (2009). Exemplification, idealization, and understanding. In Suárez (pp. 77–90).

Fodor, J. A. (1984). Semantics, Wisconsin style. *Synthese*, *59*, 231–250.

Giere, R. N. (2009). Why scientific models should not be regarded as works of fiction. In Suárez, M. (Ed.).

Godfrey-Smith, P. (1989). Misinformation. *Canadian Journal of Philosophy*, *19*(4), 533–550.

Gombrich, E. (1960). *Art and illusion. A study of psychology of pictorial representation*. London: Phaidon Press.

Goodman, N. (1972). *Problems and projects*. Indianapolis: The Bobbs-Merrill Company Inc.

Goodman, N. (1976). *Languages of art* (2nd ed.). Indianapolis: Hackett.

Hopkins, R. (1998). *Picture, image and experience*. Cambridge: Cambridge University Press.

Jones, M. R. (2005). Idealization and abstraction: A framework. In Jones & Cartwright (Eds.) (pp. 173–217).

Jones, M. R., & Cartwright, N. (Eds.). (2005). *Idealization XII–correcting the model: Idealization and abstraction in the sciences (Poznan studies in the philosophy of sciences and the humanities)* (Vol. 86). Amsterdam: Rodopi.

Kroon, F. (1988). Realism and descriptivism. In R. Nola (Ed.), *Realism and relativism in science*. Dordrecht: Kluwer.

Kulka, T. (1989). Art and science: An outline of a popperian aesthetics. *British Journal of Aesthetics*, *29*(3), 197–212.

Lewis, D. (1970). How to define theoretical terms. *Journal of Philosophy*, *62*, 427–446.

Lopes, D. (1996). *Understanding pictures*. Oxford: Clarendon Press.

Magnani, L. (2012). Scientific models are not fictions: Model-based science as epistemic warfare. In L. Magnani & P. Li (Eds.), *Philosophy and cognitive science* (pp. 1–38). Berlin: Springer.

McMullin, E. (1985). Galilean idealization. *Studies in History and Philosophy of Science*, *16*(3), 247–273.

Morrison, M. (2005). Approximating the real: The role of idealizations in physical theory. In Jones & Cartwright (Eds.) (pp. 145–172).

Niiniluoto, I. (1997). Reference invariance and truthlikeness. *Philosophy of Science*, *64*, 546–554.

Niiniluoto, I. (1999). *Critical scientific realism*. Oxford: Oxford University Press.

Niiniluoto, I. (2014). Representation and truthlikeness. *Foundations of Science*, *19*(4), 375–379.

Nola, R. (1980). Fixing the reference of theoretical terms. *Philosophy of Science*, *45*, 505–531.

Park, W. (2014). Misrepresentation in context. *Foundations of Science*, *19*(4), 363–374.

Portides, D. (2014). How scientific models differ from works of fiction. In L. Magnani (Ed.), *Model-based reasoning in science and technology* (pp. 75–87). Berlin: Springer.

Putnam, H. (1975). *Mind, language, and reality, philosophical papers* (Vol. ii). Cambridge: Cambridge University Press.

Rhodes, G. (1996). *Superportraits: Caricatures and Recognition*. East Sussex: Psychology Press.

Shelley, C. (2003). *Multiple analogies in science and philosophy*. Amsterdam: John Benjamins Publishing.

Suárez, M. (Ed.). (2009). *Fictions in science: Philosophical essays on modeling and idealization*. Routledge: New York.

Teller, P. (2009). Fictions, fictionalization, and truth in science. In M. Suárez (Ed.) (pp. 235–247).
Van Fraassen, B. C. (2008). *Scientific representation: paradoxes of perspective*. Oxford: Oxford University Press.
Woods, J. (2014). Against fictionalism. In L. Magnani (Ed.), *Model-based reasoning in science and technology* (pp. 9–42). Berlin: Springer.

Violence and Abductive Cognition

Epistemology and Ethics Entangled

Lorenzo Magnani

Abstract I think that the relationship between moral and violent behavior is still overlooked in current philosophical, epistemological, and cognitive studies. To the aim of clarifying the complex dynamics of this interplay, I will describe, adopting an eco-cognitive perspective, the concepts of salience and pregnance (originally introduced by René Thom s' catastrophe theory in semiophysical terms), and the concepts of abduction and affordance (this last one originally proposed by Gibson). Showing the interesting relationships between these four basic concepts I will explain the role of abductive cognition and affordances in building and interpreting pregnances. The main theoretical merit of the concepts of salience and pregnance is that they can be at the same time applied to physical, biological, and cognitive phenomena: it is this wide perspective which grants the possibility of presenting an integrated and systemic theory of the social role of morality and violence. Non human and human animals are endowed with internal hardwired and plastic cognitive capacities but they also continuously delegate and distribute cognitive functions to the environment to lessen their limits. Among these functions the ones devoted to produce moral frameworks in a "plastic" way are central: these activities are basically *abductive*, they create salient and pregnant moral forms, which are thought to be good to follow but that at the same time afford conflicts, from which violent outcomes can derive. The last part of this article addresses the role of pregnances as linguistic functions which are essential in building that "military intelligence" in which moral and violent behaviors, such as bullying and scapegoating, can be simply and naturally explained, in a unified perspective.

L. Magnani (✉)
Department of Philosophy and Computational Philosophy Laboratory,
University of Pavia, Pavia, Italy
e-mail: lmagnani@unipv.it

L. Magnani et al. (eds.), *Philosophy and Cognitive Science II*,
Studies in Applied Philosophy, Epistemology and Rational Ethics 20,
DOI 10.1007/978-3-319-18479-1_6

1 Abduction, Pregnances, Affordances: Eco-Cognitive Aspects

Stating that epistemology and ethics are *entangled* does not only mean that reasoning and morality can be studied together, but rather that *it benefits* to study them together. The word *entanglement* is clearly borrowed from the language of quantum physics: even if the two philosophical disciplines have each their own theoretical dignity, many of the objects they deal with are just deeply entangled, so that ignoring one aspect or the other may cause a philosophical misperception of the matter at stake. For instance, by failing to appreciate the inferential dimension in a moral judgement and its enactment, or conversely, how moral priorities strongly inform and override our best hypothetical reasonings.[1]

The entanglement of epistemology and ethics has tacitly emerged over the past recent years, transcending the philosophical impasses of the *is/ought* debate.[2] Clearing up the relationship, and the entanglement, between epistemology and ethics will help to shed light on another entangled relationship, that is the one between morality and violence. Indeed, the understanding of each theoretical entanglement (epistemology-ethics, morality-violence) rests on the understanding of the other, as the four poles are connected in a double dyadic system that I will explore in this paper.

The problem of the relationship between morality and violence can be usefully clarified taking advantage of Thom's theory of morphogenesis, based on the catastrophe theory. It is in this light that the relationship can be comprehended as an ordinary semiophysical process.[3] Furthermore, in the framework of catastrophe theory it is easy to understand the constitutive moral and at the same tome violent nature of language.[4]

To understand the basic tenets of catastrophe theory it is useful to exploit the concept of abduction, which refers to the role of "guessing hypotheses" in human and non human animal cognition. Abduction is a popular term in many fields of AI, such as diagnosis, planning, natural language processing, motivation analysis, logic programming, and probability theory. Moreover, abduction is important in the interplay between AI and philosophy, cognitive science, historical, temporal, and

---

[1]Some topics that powerfully display such entanglement are gossip studies (Bertolotti and Magnani 2014), but also any epistemological approach on religion that cannot overlook how the violence entailed by religious cognition is rooted both in the moral assumptions and in the inferential regime that are typical of religion (Bertolotti 2015), and overall the philosophical approach to the relationships between morality and violence (Magnani 2011).

[2]Such appreciation seems to be more strongly nested in applied epistemology: David Coady explicitly connects the origins of applied epistemology to the tradition of applied ethics (Coady 2012, p. 1 and ff.), highlighting a theoretical practice of mutual borrowing that has characterized the different branches of philosophy since the very beginning.

[3]Thom considered the use of models in catastrophe theory as illustrating semiophysical processes, which in the case of cognition express what he called a "physics of meaning" (Thom 1988, Foreword).

[4]On the violent nature of language in a philosophical, perspective see Magnani (2011, Chap. 1).

narrative reasoning, decision-making, legal reasoning, and emotional cognition.[5] Eight volumes (monographs and collections) are currently available (Josephson and Josephson 1994; Flach and Kakas 2000; Kuipers 2000; Magnani 2001; Gabbay and Woods 2005; Aliseda 2006; Walton 2004; Magnani 2009) and three special issues of international journals (*Philosophica*, 1998 61(1); *Foundations of Science*, 2004, 9; 2008, 13(1); *Logic Journal of the IGPL*, 2006 14(1)). Of course many articles from various disciplinary fields of research are continually published on this topic.[6] To illustrate the concept of abduction let us consider the following interesting passage, from an article by Simon (1965), dealing with the logic of normative theories:

> The problem-solving process is not a process of "deducing" one set of imperatives (the performance programme) from another set (the goals). Instead, it is a process of selective trial and error, using heuristic rules derived from previous experience, that is sometimes successful in *discovering* means that are more or less efficacious in attaining some end. If we want a name for it, we can appropriately use the name coined by Peirce and revived recently by Norwood Hanson (1958): it is a *retroductive* process. The nature of this process – which has been sketched roughly here – is the main subject of the theory of problem-solving in both its positive and normative versions (Simon 1977, p. 151).

Simon states that discovering means that are more or less efficacious in attaining some end are performed by a *retroductive* process. He goes on to show that it is easy to obtain one set of imperatives from another set by processes of discovery or retroduction, and that the relation between the initial set and the derived set is not a relation of logical implication. I completely agree with Simon: retroduction (that is abduction, cf. below) is the main subject of the theory of problem-solving and developments in the fields of cognitive science and artificial intelligence have strengthened this conviction.

Hanson (1958, p. 54) is perfectly aware of the fact that an enormous range of explanations (and causes) exists for any event:

> There are as many causes of x as there are explanations of x. Consider how the cause of death might have been set out by a physician as "multiple hemorrhage", by the barrister as "negligence on the part of the driver", by a carriage-builder as "a defect in the brakeblock construction", by a civic planner as "the presence of tall shrubbery at that turning".

The word "retroduction" used by Simon is the Hansonian neopositivistic one replacing the Peircean classical word abduction. Following Hanson's point of view Peirce "[...] regards an abductive inference (such as 'The observed position of Mars falls between a circle and an oval, so the orbit must be an ellipse') and a perceptual judgment (such as 'It is laevorotatory') as being opposite sides of the same coin". It is also well-known that Hanson relates abduction to the role of patterns in reasoning and to the Wittgensteinian "Seeing that" (Hanson 1958, p. 86).

As Fetzer has stressed, from a philosophical point of view the main modes of argumentation for reasoning from premises to conclusions are expressed by these

---

[5]A list of the classical bibliography on abduction is given in Magnani (2001).

[6]General considerations on the basic aspects of abduction in science and AI can also be found in Gooding (1996); Josephson and Josephson (1994); Kuipers (1999); Thagard (1988); Shrager and Langley (1990).

Fig. 1 Creative and selective abduction

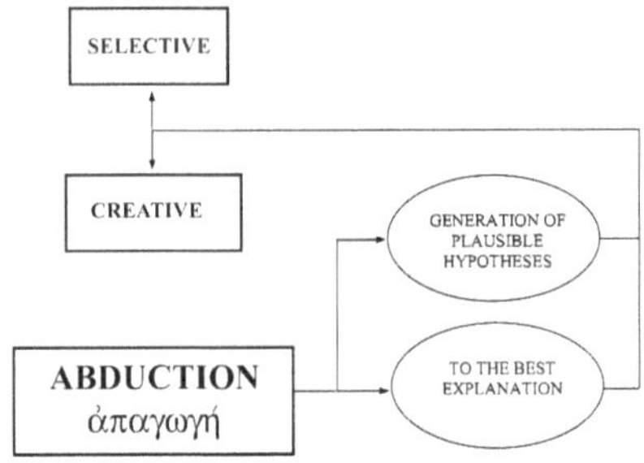

three general kinds of reasoning: *deductive* (demonstrative, non ampliative, additive), *inductive* (non-demonstrative, ampliative, non additive), *fallacious* (neither, irrelevant, ambiguous). Abduction, which expresses likelihood in reasoning, is a typical form of fallacious inference (at least in the perspective of classical logic): "[...] it is a matter of utilizing the principle of maximum likelihood in order to formalize a pattern of reasoning known as 'inference to the best explanation' " (Fetzer 1990, p. 103).[7]

To conclude this short digression on abduction it is worth to recall the distinction[8] between two kinds of abduction, *theoretical* and *manipulative*. The first one mainly takes advantage of internal cognitive resources, the second also and primarily exploits all kinds of external representations, cognitive mediators, and cognitive artifacts (consider for instance the use of epistemic mediators in scientific practice, such as computational representations or *in vitro* models). In particular, manipulative abduction shows how we can find methods of manipulative constructivity, to the aim of making hypotheses.

In both cases various kinds of representations can work, from the *model-based* representations (for example icons, diagrams, spatial frameworks, etc.) to the *sentential* ones. Still in both cases two main cognitive aspects of abduction can be found: (1) abduction that only generates "plausible"[9] hypotheses ("selective" or "creative") and (2) abduction considered as inference "to the best explanation", which also evaluates hypotheses (cf. Fig. 1) (Harman 1973; Thagard 1988; Lipton 2004). An illustration of creative abduction from the field of medical knowledge is represented by the discovery of a new disease and the manifestations it causes. Therefore, "creative" abduction deals with the whole field of the growth of scientific knowledge. This is irrelevant in medical diagnosis where instead the task is to abductively "select" from an encyclopedia of pre-stored diagnostic entities. We can call both inferences

[7] On the inference to the best explanation see also Harman (1965, 1968), Thagard (1987), Lipton (2004).

[8] Further illustrated in Magnani (2009), I introduced this distinction in Magnani (2001). The distinction between creative and selective abduction illustrated below, was introduced in an article of 1988 (Magnani 1988).

[9] A further analysis of this important concept is illustrated in Section Magnani (2009, Chap. 2).

ampliative, selective and creative, because in both cases the reasoning involved amplifies, or goes beyond, the information incorporated in the premises (Magnani 1992).

Taking advantage of the concept of abduction first of all we need clarify in the following paragraphs the notions of pregnance and salience, which play an important role in the catastrophe theory. An example of a special case of abduction, instinctual (and putatively "unconscious") is given by the case of certain cognitive abilities embodied in animals. These abilities are in turn capable of leading to some appropriate behavior: as Peirce said, abduction even takes place when a new born chick picks up the right sort of corn. Following Peirce, I have contended (Magnani 2009, Chap. 5) this is an example of spontaneous abduction—analogous to the case of other hardwired unconscious/embodied abductive processes in human beings:

> When a chicken first emerges from the shell, it does not try fifty random ways of appeasing its hunger, but within five minutes is picking up food, choosing as it picks, and picking what it aims to pick. That is not reasoning, because it is not done deliberately; but in every respect but that, it is just like abductive inference.[10]

It is clear that Peirce considers hypothesis generation a largely instinctual endowment[11] of human beings given by God or related to a kind of Galilean "*lume naturale*": "It is a primary *hypothesis* underlying all abduction that the human mind is akin to the truth in the sense that in a finite number of guesses it will light upon the correct hypothesis" (Peirce CP, 7.220). Hence, human mind is "akin to truth", but this tendency is also present in animals. Again, the example of the innate ideas of "every little chicken" is of help to describe this human instinctual endowment:

> How was it that man was ever led to entertain that true theory? You cannot say that it happened by chance, because the possible theories, if not strictly innumerable, at any rate exceed a trillion – or the third power of a million; and therefore the chances are too overwhelmingly against the single true theory in the twenty or thirty thousand years during which man has been a thinking animal, ever having come into any man's head. Besides, you cannot seriously think that every little chicken, that is hatched, has to rummage through all possible theories until it lights upon the good idea of picking up something and eating it. On the contrary, you think the chicken has an innate idea of doing this; that is to say, that it can think of this, but has no faculty of thinking anything else. The chicken you say pecks by instinct. But if you are going to think every poor chicken endowed with an innate tendency toward a positive truth, why should you think that to man alone this gift is denied? (Peirce CP, 5.591)

I think the concept of pregnance, introduced by Thom (1972, 1980) on the basis of Wertheimer's Gestaltic concept of *Prägnanz*, can shed further light on a kind of morphodynamical "physics" of abduction, first of all in the case of the instinctual hardwired aspects I have just illustrated. Furthermore, pregnance and salience can became clearer and richer when reframed in the perspective of abductive cognition. As I will soon show, they are key concepts which can be exploited to illustrate

[10]Cf. the article "The proper treatment of hypotheses: a preliminary chapter, toward an examination of Hume's argument against miracles, in its logic and in its history" [1901] (in Peirce 1966, p. 692).

[11]Instinct is of course in part conscious: it is "always partially controlled by the deliberate exercise of imagination and reflection" (Peirce CP, 7.381).

important aspects not only of the instinctual but also of the plastic nature of abductive hypothetical cognition.[12]

As they acquire their meaning in a very naturalistic intellectual framework related to an analysis of complex systems, they are also useful to propose a unified perspective on moral and violent abductive processes and other various cognitive/psychic ones, seen as basic physico-biological events, also endowed with a profound eco-cognitive significance. The pregnance affects an organism, and the related abductive/hypothetical many kinds of responses are promptly triggered, biological, physical, cognitive. Here it is important to stress that in gregarious animals the triggered response is often a proto-moral/proto-violent one.[13] Hence, pregnances are genetically transmitted but can also be actively and plastically created for example through learning and high cognitive capacities, through the formation of multiple forms of hypothetical intelligence.

1.1 Saliences and Pregnances as Biological and Cognitive Mediators

What is a pregnance? The complicated—and at first sight obscure—concept of *pregnance* is based on the concept of *salience*, which emerges in the dynamical framework of the "semiophysical" perspective. First of all we can say that in general, phenomenal discontinuities are perceived by organisms as *salient forms* (for example, in the auditive case, the eruption of a sound in the midst of silence), that is, as contextual effects between forms: "The simplest feature is the punctual discontinuity geometrically represented by a point dividing the real straight line **R** into two half lines" (Thom 1988, p. 3). Discontinuities *out there* in the environment are basically *translated* into other more or less amplified discontinuities in the subjective sensorial state, as a kind of "echo" or "shock" of the physical environment *within* an organism. In the case of sensory systems, salience of course is at the basis of the first possibility of perceiving individuated forms. In this case perception can also be appropriately influenced by a certain form of *concept* "[...] that is to say a class of equivalence between forms referent to the same concept": the lack of the concept can annihilate the grasping of the individuated form, especially when analysis proceeds from the whole to the parts.

To the aim of the present study, it is important to stress that the term pregnance can be applied to physical and biological phenomena, but also to the cognitive ones. Hence, it can further clarify the distinction between the instinctual chicken abduction above and other plastically acquired abductive ways of cognition:

---

[12]The plastic nature of abductive cognition refers to all the skillful capacities to make hypotheses, which human beings are able to learn and exploit.

[13]Some non-human animal behaviors can reasonably be called proto-moral, to lessen the anthropomorphic aura of the adjective "moral" (Waal et al. 2006).

> So we will get this general pattern of a world made up of salient forms and pregnances – salient forms being objects, very often individuated, that are impenetrable to one another, and pregnances being occult qualities, efficient virtues that emanate from source-forms and invest other salient forms in which they produce visible effects (that is the so-called "figurative" effects for the organisms invested) (Thom 1988, p. 2).

Let us explain the passage. First of all we have to note that when Thom calls pregnances "occult qualities", that is just a metaphor: actually Thom thinks that pregnances are not occult and mysterious qualities at all, because they could be accounted for as fully explainable psychological phenomena in neurological and biological terms, and they can also be made intelligible through mathematical models. The description of the processes affected by pregnance activity aims at providing what Thom calls a "protophysics, source and reservoir of all permanent intuitions, of all those archetypal metaphors that have nourished man's imagination over the ages" (p. 3).

Thom further says: "*Pregnances* are non-localized entities emitted and received by salient forms. When a salient form 'seizes' a pregnance, it is invaded by this pregnance and consequently undergoes transformations in its inner state which can in turn produce outward manifestations in its form: we call these *figurative effects*" (p. 16). To clarify the two concepts of salience and pregnance the following two examples can be of some utility [the wide range of events covered by the two concepts is testified by the fact that the first example does not have any cognitive/psychological significance]: (1) an infection (pregnance) contaminates healthy subjects (representing the "invested" form: salience). These subjects in turn re-emit the same infection (pregnance) into the environment. In this case pregnance has in itself a material/biological support (for example a virus)—as a mediator, which in turn is transmitted thanks to a suitable medium (for example air or blood); (2) worker honeybees communicate with each other by means of signs (through the iconic movements of a dance)—pregnance—that express the site where they have found food in order to inform the other conspecific individuals—the invested salience—about the location. In this second case the pregnance is transmitted—mediated—through undulatory sounds and light signals and produces a neurobiological effect at the destination, that is, in other words, a "psychic" effect [of course we can use in this case the expression "psychic" only if we admit, in a mentalistic and unorthodox way, that honeybees are endowed with *a kind of* animal psyche: an example regarding a cat or a boy would have been more convincing for the reader…].

Finally, fields in physics are the true paradigm of *objective pregnances* in modern science, because in that case we are theoretically able to calculate their variation in space-time thanks to a mathematical description (based on an explicit geometrical definition of space-time) (Thom 1988, p. 32). However, to better grasp the concept of pregnance and its relationship with moral and violent social behaviors a further analysis is needed.

1.2 *Eco-Cognition of Moral Pregnances and Affordances*

In general, in the case of salient forms, their impact on the organism's sensory apparatus "remains transient and short lived" (Thom 1988, p. 2), so they do not have relevant long-term effects on the behavior of the organisms. To continue and deepen our analysis, it is useful at this point to introduce the concept of *affordance*. If we acknowledge that environments and organisms evolve and change, and also both their instinctual and cognitive plastic endowments, we may argue that affordances (Gibson 1951, 1979, 1982) can be related to the variable (degree of) "abducibility" of a configuration of signs[14]: a chair affords sitting in the sense that the action of sitting is a result of a sign activity in which we perceive some physical properties (flatness, rigidity, etc.), and therefore we can ordinarily infer that a possible way to cope with a chair is sitting on it.[15]

In the case of cognitive events, if we adopt the perspective of the affordances, we can say that salient forms—contrary to pregnant forms, "afford" organisms without triggering *relevant* modifications either at the level of possible inner rumination or in terms of motor actions. Thom says that when salient forms carry "biological significance", like in the *form of prey* for the hungry predator, or the predator for its prey, or in the case of sex and fear, or when a salient form is invested by an infection, the reaction is much bigger and involves the freeing of hormones, emotive excitement and behavior (or an immune response in the case of the infection) devoted to attracting or repulsing the form: *salient forms of this type are called pregnant*.

However, in the perspective of the complexity of animal behaviors, in the non strictly biological case of *cognitive* functions we still have to deal with pregnances. In this case, *pregnances*, no matter whether due to innate releasing processes or to complicated, more or less stable internal learnt processes and representations (or pseudorepresentations (Bermúdez 2003), such is the case of non-human animals) are triggered by a very small sensory stimulus (a stimulus "with a little figuration, an olfactory stimulus for instance" (p. 6)). Hence, they represent a relationship with certain *special* phenomenological aspects, that of course are stable to different extents and so can appear and disappear. At some times and in some cases the special sensitivity to pregnances is disregarded. Like in the case of affordances, this variability and transience can be seen at the level of the differences of pregnance sensitivity among organisms and also at the level of the same organism at subsequent stages of its cognitive and biological development. We can say that a pregnant stimulus is—so to say—*highly diagnostic* and a trigger to initiate abductive cognition, like in the case of the hardwired pregnance occurring to our Peircean chicken and its food: the chicken promptly reacts when perceiving it. When a pregnance affects an organism, the abductive reaction can be promptly triggered. It has to be said that in this case what can be surely seen as a biological/istinctual reaction—reaching the food—it is at the same time endowed with a kind of "compacted" cognitive value, like Peirce

[14]In this case we adopt the semiotic/Peircean lexicon which refers to cognition as sign activity.

[15]Cf. Magnani (2009, Chap. 6).

brilliantly contends: non-human animal mind is already "akin to truth": indeed, in this regard, let us reiterate the passage already reported above

> [...] you think the chicken has an innate idea of doing this; that is to say, that it can think of this, but has no faculty of thinking anything else. The chicken you say pecks by instinct. But if you are going to think every poor chicken endowed with an innate tendency toward a positive truth, why should you think that to man alone this gift is denied? (Peirce CP, 5.591)

Finally, we have to recall that the pregnant character of a form is always relative to a receiving subject (or group of subjects), just as in the eco-psychological case of affordances.

Pregnances can be abductively activated or created. When a bell ringing is repeated often enough together with the exhibition of a piece of meat to a dog, thanks to Pavlovian conditioning the alimentary pregnance of meat spreads by contiguity to the salient auditive form, so that the salient form, in this case the sound of the bell, is invested by the alimentary pregnance of the meat; here the metaphor of the invasive fluid—even if exoteric—can be useful: "So we can look on a pregnance as an invasive fluid spreading through the field of perceived salient forms, the salient form acting as a 'fissure' in reality through which seeps the infiltrating fluid of pregnance" (p. 7).[16] The propagation can also occur through similarity, taking advantage of the mirroring force of some features. Once the reinforcement is established, the bell—Thom says—refers *symbolically* in a more or less stable way, to the meat. In these cases we can say, metaphorically and anthropomorphically, we are facing with an example of "emergence of meaning".

Of course extinction of pregnances through absence of reinforcement is possible, when an organism moves away for a long time from the source form or when the invested salient form is associated with another pregnant form still in absence of reinforcement. From this point of view the "symbolic activity" seen in the above situation is seen as fundamentally linked to biological control systems in two ways: (1) it is an extension of their efficacy (new favorable cognitive abductive chances—new pregnances—are added); (2) an internal simulation concerning the relationship between the food and its index, the bell, is implemented, so that the door is opened to the formation of multiple forms of abductive semiotic cognition (and /or intelligence):

> The fact that initially, as in the Pavlovian schema, this stimulation is no more than a simple association, does not stop us from considering that we have the first tremors in the plastic and competent dynamic of the psychism of [the actant] of an external spatiotemporal liaison interpreted not without reason as causal. [...] Hence, from the beginning, the situation is not fundamentally different from that of language [...]. Only these fundamental "catastrophes" of biological finality have the power of generating the symbols in animals (Thom 1988, pp. 268–269).

[16]To explain the formation of pregnances Thom exploits the classical Pavlovian perspective. More recent approaches take advantage of Hebbian (Hebb 1949) and other more adequate learning principles and models, cf. for example Loula et al. (2010).

1.3 The Artlessness of Proto-Morality and Violence

As I have already indicated Thom sees pregnances not only as innate endowments (like in the case of the basic ones seen in birds and mammals: hunger, fear, sexual desire), but also as related to higher-level cognitive capacities, which also involve the role of *proto-morality* (in non human animals) and *morality* (in beings-like-us). "When animal pregnance is generalized in the direction of human conceptualization 'conceptual' or individuating pregnances will be revealed, the nature of which is close to 'salience'" (p. 6). At this point it should be clear that I maintain we can synthetically account for both these processes in terms of different kinds of abductive hypothetical cognition. For example, Thom observes, reverberating the view of visual perception as semi-encapsulated,[17] that "[...] it is doubtful whether genetics alone would be able to code a *visual* form [...]. Whence the necessity of invoking cultural transmission, linked with the social or family organization of the community" (p. 10). In gregarious animals the signals (which also have to be seen as referring to the explanation of the origin of the "pregnance-mirroring" functions of human language) are a vector of pregnances insofar as they transfer a pregnance from one individual to another, or to several others. In such a way they favor teaching and learning, working to constitute the collectively shared behavior needed for example to capture food and to ward off predators. In this perspective of gregarious animals pregnances are de facto immediately related to the emergence of kinds of proto-moralities relying on shared proto-axiological features.[18]

When an organism—through abductive cognition[19]—traces back a symbolic reference to a "source" form [in Thom's sense as indicated above], often a motor reaction becomes necessary to bring satisfaction. Here an example that is clearly and patently related to "sociality" (through morality, given the presence of the role of altruism):

> In a social group, one individual's encounter with a source form *S* may give rise to a dilemma: whether to pursue the "individual interest" which consists in using the regulatory reflex that will result in selfish satisfaction, or to follow the altruistic community strategy by uttering the cry that will carry the pregnance *S* to the other members of the community; such a cry is then the signal by which the signal *P* of *S* experienced by individual 1 can be transferred to another individual 2 (p. 12).

Thom himself nicely adds that this kind of animal proto-moral conflict resonates with the more clearly "moral" conflict of civilized societies "This dilemma exists well

---

[17]Perception is informationally "semi-encapsulated", and also pre-wired, i.e., despite its bottom-up character, it is not insulated from plastic cognitive processes and contents acquired through learning and experience, cf. Raftopoulos (2009).

[18]The emergence of proto-morality and proto-violence can also be naturalistically seen in an evolutionary perspective, as I have illustrated in Magnani (2011, Chap. 1).

[19]As already illustrated above, in this case abduction plays and inferential role similar to the one it plays in physician's *diagnostic* reasoning, when a symptom is explained by a hypothesis, a diagnosis, suitably selected among an already available encyclopedia of diagnostic hypotheses referred to the corresponding diseases. On the contrary, when a pregnancy is originally built, the process is akin to the case of *creative* abductive cognition, for example in science, when a new successful hypothesis is established for the first time. On these aspects of abductive cognition see Magnani (2009, Chap. 2).

and truly in our society. Witness the scruples most honest citizens have in making true declarations of their taxable revenues" (Thom 1988, p. 12).

An example is provided by the case of a signal (or a proximal "clue"), which transfers the pregnance of fear in birds, which further prompts the motion of taking flight but that also incurs the risk of attracting the predator's attention. Animals perceive the pregnant sign/clue (for example tracks or excreta of the predator), and then emit a further sign (cry) that mirrors that sign/clue and its pregnance.

At this point it is clear that, in this Thomian perpective, the establishment of a proto-morality immediately depicts behaviors and reactions that are exposed to punishment and violence: at the ame time moral behavior creates the space of violent behavior.

2 The Moral/Violent Function of Language

From the point of view of the functions of human language Thom sees the birth of the "genitive" as the syntactical form that denotes the proximity of a being whilst denying its immediate presence. This syntactical form permits us to emit and receive alarm calls[20] which provide individuals (and the group) with an adequate defense. From this perspective the presence of a pregnant sign associated with a form *S* can be considered as a fundamental kind of *concept* or class of equivalence between salient forms, which incorporates a primary, rudimentary and prompt abductive power.

As I have already stressed, the *cultural* acquisition of a sensitivity to source forms has to be hypothesized in both humans and various animals. In these cases pregnance transmission occurs, beyond the hardwired cases, thanks to the presence of suitable artifactual cognitive niches[21] (such as human natural languages), functioning as *pregnance mediators*, where plastic teaching and learning is possible. These cognitive niches make plenty of cognitive tools available, that in turn make the organisms who acquire them able to *pregnantly* manage signs (which consequently gain a special "meaning"). This process is clearly illustrated by the description of various aspects of "plastic"—and not merely hardwired—cognitive skills in animal abduction and by the relevance of the "mediated" character of several affordances. In these last two cases both cognitive skills and sensitivity to suitable affordances require cultural learning/training imbued in appropriate cognitive niches.

In Chap. 5 of my book on abductive cognition[22] I have emphasized that fleeting and evanescent internal pseudorepresentations (beyond reflex-based innate releasing processes, trial and error or mere reinforcement learning) are needed to account for many animal "communication" performances even at the rudimentary level of

[20] Also, in many animals alarm calls/cries are the analogue of the second-person singular imperatives typical of human natural languages (Thom 1980, p. 172).

[21] This concept, introduced by Tooby and DeVore (1987), and later on reused by Pinker (1997, 2003), is illustrated in Magnani (2009, Chap. 5).

[22] Magnani (2009).

chicken calls: Evans says that "[...] chicken calls produce effects by evoking representations of a class of eliciting events [food, predators, and presence of the appropriate receiver] [...]. The humble and much maligned chicken thus has a remarkably sophisticated system. Its calls denote at least three classes of external objects. They are not involuntary exclamations, but are produced under particular social circumstances" (Evans 2002, p. 321): in Thom's words, these calls are of course pregnant signals which can be learnt, which in turn play a proto-moral and a kind of "deontological" role by triggering reactions that are implicitly considered good. Of course in the case of animal cheating, analogous calls trigger reactions that are basically negative for the receiver's welfare.[23]

Chickens form separate representations when faced with different events and they are affected by prior experience (of food, for example). These representations are mainly due to internally developed plastic capacities to react to the environment, and can be thought of as the fruit of learning. Many animals (especially gregarious ones) go beyond the use of sound signals in their cognitive performances, they for example *reify* and delegate cognitive/semiotic roles to true pregnant external artificial "pseudorepresentations" (for example landmarks, urine-marks, etc.) which artificially modify the environment to consequently become an affordance for themselves and other individuals of the group or of other species.

2.1 "Military Intelligence", Morality, and Ideologies

Taking advantage of the conceptual framework brought up by Thom's catastrophe theory on how natural syntactical language is seen as the fruit of social necessity, its fundamental function can only be clearly seen if linked to an intrinsic *moral* (and at the same time *violent*) aim which is basically rooted in a kind of *military intelligence*, which relates to the problem of the role of language in the so-called *coalition enforcement* (Bingham 1999, 2000), that is in the affirmation of morality and the related perpetration of violent punishment. To anticipate the content of this section taking advantage of a kind of motto, I can say: "when words distribute moral norms and habits, often they also wound and inflict harm".

Thom says language can simply and efficiently transmit *vital* pieces of information about the fundamental biological oppositions (life—death, good—bad): it is from this perspective that we can clearly see how human language—even at the level of more complicated syntactical expressions—always carries information (pregnances) about moral qualities of persons, things, and events. Such qualities are always directly or indirectly related to the survival needs of the individual and/or of the group/coalition.

[23]Deception in animals is synthetically illustrated in El-Hani et al. (2009).

The syntactical aspects of natural languages,[24] such as for example the genitive—I have already quoted above—or the management of the verb,[25] can be further explained in terms of archetypal morphological space-time processes susceptible of being described in mathematical topological terms. For instance "Upon hearing an order the cerebral dynamics suffers a specific stimulus s, which sends it into an unstable state of excitation. This state then evolves towards stability through its capture by the attractor A, whose excitation generates the motor execution of the order by coupling the motor neurones" (Thom 1980, p. 172). For example, the verbs of feeling *to fear* and *to hope* express that an actant subject admits in internal co-ordinates "a morphology of the future, which is accepted with repulsion or attraction" (p. 211).

Thom too is convinced of the important role played by language in maintaining the structure of societies, defending it thanks to its moral and violent role: "information has a useful role in the stability or 'regulation' of the social group, that is, in its defence" (Thom 1988, p. 279). When illustrating "military" and "fluid" societies he concludes:

> In a military type of society, the social stability is assured, in principle, by the imitation of the movement of the hierarchical superior. Here it is a question of a slow mechanism where the constraints of vital competition can impose rapid manoeuvres on the group. Also the chief cannot see everything and has need of special informers stationed at the front of the group who convey to him useful information on the environment. The invention of a sonorous language able to communicate information and to issue direction to the members of the group, has enabled a much more rapid execution of the indispensable manoeuvres. By this means (it is not the only motivation of language), one can see in the acquisition of this function a considerable amelioration of the stability of a social group.
>
> If language has been substituted for imitation, we should note that the latter continues to play an important role in our societies at pre-verbal levels (cf. fashion). In addition, imitation certainly plays a primary part in the language learning of a child of 1 to 3 years (pp. 235–236).

I have illustrated in Magnani (2011, Chap. 1), taking advantage of both an evolutionary and a paleoanthropological cognitive perspective, that in human or pre-human groups the appearance of coalitions dominated by a central leader quickly leads to the need for surveillance of surrounding territory to monitor prey and free-riders and watch for enemies who might jeopardize the survival of the coalition.

This is an idea shared by Thom who believes that language becomes a fundamental tool for granting stability and favoring the indispensable manipulation of the world "thus the localization of external facts appeared as an essential part of social communication" (Thom 1988, p. 26), a performance that is already realized by naming[26] (the containing relationship) in divalent structures: "X is in Y is a basic form of investment (the localizing pregnance of Y invests X). When X is invested with a

---

[24]Basic syntactical mechanisms are intended by Thom as simulated copies (defined on an abstract space) of the fundamental biological functions such as predation and sexuality.

[25]For example a verb transfers a pregnance from subject to object and so constitutes an attractor of the cerebral dynamics.

[26]It is important to stress that pregnant forms, as they receive names, tend to loose their alienating character.

ubiquitous biological quality (favorable or hostile), then so is *Y*" (ibid.). A divalent syntactical structure of language becomes fundamental if a *conflict* between two outside agents has to be reported. The trivalent syntactical structure subject/verb/object forges a salient "messenger" form that conveys the pregnance between subject and recipient. In sum, the usual abstract functions of syntactic languages, such as conceptualization, appear strictly intertwined with the basic social and especially *military* nature of communication.

2.2 Language and Conflicts

I contend that this military nature of linguistic communication is intrinsically "moral" (protecting the group by obeying shared norms), and at the same time "violent" (for example, killing or mobbing to protect the group). This basic moral/violent effect can be traced back to past ages, but also when we witness a somehow *un-civil* use of everyday natural language in current mobbers, who express strategic linguistic communications "against" the mobbed target. These strategic linguistic communications are often performed thanks to hypothetical reasoning, abductive or not. In this case the use of natural language can take advantage of efficient hypothetical cognition through gossip, fallacies and so on, but also of the moral/violent exploitation of apparently more respectable and sound truth-preserving and "rational" inferences. The narratives used in a dialectic and rhetorical setting qualify the mobbed individual and its behavior in a way that is usually thought of by the mobbers themselves (and by the individuals of their coalition/group) as moral, neutral, objective, and justified while at the same time hurting the mobbed individual in various ways. Violence is very often subjectively dissimulated and paradoxically considered as the act of performing just, objective moral judgments and of persecuting moral targets. In sum, *de facto* the mobbers' coordinated narratives harm the target (just as if she were being *stoned* in a ritual killing), very often without an appreciable awareness of the violence performed.

This human linguistic behavior is clearly made intelligible when we analogously see it as echoing the anti-predatory behavior which "weaker" groups of animals (birds, for example) perform, for example through the use of suitable alarm calls and aggressive threats. Of course such behavior is mediated in humans through socially available ideologies (differently endowed with moral ideas) and cultural systems. Ideologies can be seen as fuzzy and ill-defined cultural mediators spreading pregnances that invest all those who put their faith in them and stabilize and reinforce the coalitions/groups: "[...] the follower who invokes them at every turn (and even out of turn) is demonstrating his allegiance to an ideology. After successful uses the ideological concepts are extended, stretched, even abused", so that their meaning slowly changes in imprecise (and "ambiguous", Thom says)[27] ways, as we have seen

[27] From this perspective the massive moral/violent exploitation of equivocal fallacies in ideological discussions, oratories, and speeches is obvious and clearly explainable.

happens in the application of the archetypical principles of mobbing behavior. That part of the individual unconscious we share with other human beings—i.e. a kind of Jungian *collective unconscious*—shaped by evolution—contains archetypes like the "scapegoat" (mobbing) mechanism I have already mentioned.

In this cognitive mechanism, a paroxysm of violence focuses on an arbitrary sacrificial victim and a unanimous antipathy would, mimetically, grow against him. The process leading to the ultimate bloody violence (which was, for example, widespread in ancient and barbarian societies) is mainly carried out in current social groups through linguistic communication. Following Girard (1977, 1986) we can say that in the case of ancient social groups the extreme brutal elimination of the victim would reduce the appetite for violence that had possessed everyone just a moment before, leaving the group suddenly appeased and calm, thus achieving equilibrium in the related social organization (a sacrifice-oriented social organization may be repugnant to us but is no less "social" just because of its rudimentary violence).

This kind of archaic brutal behavior is still present in civilized human conduct in rich countries and is almost always implicit and unconscious, for example in that racist and mobbing behavior I have already quoted. Let me reiterate that, given the fact that this kind of behavior is widespread and partially unconsciously performed, it is easy to understand how it can be implicitly "learned" in infancy and still implicitly "pre-wired" in an individual's cultural unconscious (in the form of ideology as well) that we share with others as human beings. I strongly believe that the analysis of this archaic mechanism (and of other similar *moral/ideological/violent* mechanisms) might shed new light on what I call the basic equivalence between engagement in morality and engagement in violence since these engagements, amazingly enough, are almost always hidden from the awareness of the human agents that are actually involved.

Recent evolutionary perspectives on human behavior, taking advantage of neuroscience and genetics[28] have also illustrated the related process of *otherisation*—which decisively primes people for aggression—as a process grounded in basic human emotions, i.e. our bias towards pleasure and avoidance of pain. Perceiving others as the "others" causes fear, anger or disgust, universal "basic" responses to threats whose physiological mechanisms are relatively well understood. It is hypothesized that these emotions evolved to enable our ancestors to escape predators and fight enemies. Of course the otherisation process continues when structured in "moral" terms, like for example in the construction of that special other that becomes a potential or actual scapegoat.

---

[28]Taylor (2009). Taylor's book also provides neuroscientific explanations on how brains process emotions, evoke associations, and stimulate reactions, which offer interesting data—at least in terms of neurological correlates—on why it is reactively easy for people to harm other people.

2.3 Scapegoating Through Pregnances

It is worth mentioning, in conclusion, the way Thom accounts for the social/moral phenomenon of scapegoating in terms of the complexity of pregnances. "Mimetic desire", in which Girard (1986) roots the violent and aggressive behavior (and the scapegoat mechanism) of human beings can be seen as the act of appropriating a desired object which imbues that object with a pregnance, "the same pregnance as that which is associated with the act by which 'satisfaction' is obtained" (Thom 1988, p. 38). Of course this pregnance can be propagated by imitation through the mere sight of "superior" individuals[29] in which it is manifest: "In a sense, the pleasure derived from looking forward to a satisfaction can surpass that obtained from the satisfaction itself. This would have been able to seduce societies century after century (their pragmatic failure in real terms having allowed them to escape the indifference that goes with satiety as well as the ordeal of actual existence)" (ibid.).

Recent cognitive research stresses the influence that *intentional gaze* processing has on "object processing": objects falling under the gaze of others acquire properties that they would not display if not looked at. Observing another person gazing at an object enriches it of motor, affective, and status properties that go beyond its chemical or physical structure. We can conclude that gaze plays the role of transferring to the object the intentionality of the person who is looking at it. This result further explains why mimetic desire can spread so quickly among people belonging to specific groups.[30]

Grounded in appropriate wired bases, "mimetic desire" is indeed a sophisticated template of behavior that can be picked up from various appropriate cultural systems, available over there, as part of the external cognitive niches built by many human collectives and gradually externalized over the centuries (and always transmitted through activities, explicit or implicit, of teaching and learning), as fruitful ways of favoring social control over coalitions. Indeed mimetic desire triggers envy and violence but at the same time the perpetrated violence causes a reduction in appetite for violence, leaving the group suddenly appeased and calm, thus achieving equilibrium in the related social organization through a *moral effect*, that can in turn become a possible *carrier* of further violence.

Mimetic desire is related to envy (even if of course not all mimetic desire is envy, certainly all envy is mimetic desire): when we are attracted to something the others have but that we cannot acquire because others already possess it (for example because they are rival goods), we experience an offense which generates envy. In the perspective introduced by Girard envy is a mismanagement of desire and it is of capital importance for the moral life of both communities and individuals. As a reaction to offense, envy easily causes violent behavior. From this point of view we can psychoanalytically add that "[...] the opposite of egotist self-love is not altruism, a concern for common good, but envy and resentment, which makes me act against

[29] Or through the exposure to descriptions and narratives about them and their achievements.

[30] Becchio et al. (2008). On gaze cueing of attention cf. also Frischen et al. (2007), who also established that in humans prolonged eye contact can be perceived as aggressive.

my own interests. Freud knew it well: the death drive is opposed to the pleasure principle as well as to the reality principle. The true evil, which is the death drive, involves self-sabotage" (Žižek 2009, p. 76).

3 Conclusion and Future Work

In this paper I have illustrated, taking advantage of the concepts of salience and pregnance, derived from the catastrophe theory, and of the concept of abduction and affordance, some eco-cognitive aspects of moral and violent behavior. I have also offered new insight on the analysis of the strict relationships between these behaviors, by offering a unified perspective rooted in a morphodynamical framework in which physical, biological, and cognitive processes can be simultaneously analyzed. The approach described in this paper promises new developments in many directions, beyond merely socio-moral aspects. For example, the analysis of finance as a semiocognitive and at the same time as an artifactual hyper-technological niche, which imposes itself over market competition, but which cannot make the necessary gains from market competition (which conversely it impairs), opens up the analysis of the violent related outcomes. Indeed, if seen not only as a technological but also as a semiocognitive niche, current global financial systems in many cases appear to be carriers of prophecies that aim at being self-fulfilling, but fall short of it because prophets are not even human but cyborgs or artificial intelligences: violent effect against humans derive.[31]

References

Aliseda, A. (2006). *Abductive reasoning. Logical investigations into discovery and explanation*. Berlin: Springer.

Becchio, C., Bertone, C., & Castiello, U. (2008). How the gaze of others influences object processing. *Trends in Cognitive Science, 12*(7), 254–258.

Bermúdez, J. L. (2003). *Thinking without words*. Oxford: Oxford University Press.

Bertolotti, T. (2015). *Patterns of rationality: Recurring inferences in science, social cognition and religious thinking*. Berlin/Heidelberg: Springer.

Bertolotti, T., & Magnani, L. (2013). Terminator niches. In *Proceedings of the Virtual Reality International Conference: Laval Virtual, VRIC'13* (pp. 31:1–31:10), New York: ACM Digital Library.

Bertolotti, T., & Magnani, L. (2014). An epistemological analysis of gossip and gossip-based knowledge. *Synthese, 191*, 4037–4067.

Bingham, P. M. (1999). Human uniqueness: A general theory. *The Quarterly Review of Biology, 74*(2), 133–169.

Bingham, P. M. (2000). Human evolution and human history: A complete theory. *Evolutionary Anthropology, 9*(6), 248–257.

[31] Some preliminary suggestions concerning the analysis of economical systems as semiocognitive niches is provided in Bertolotti and Magnani (2013).

Coady, D. (2012). *What to believe now: Applying epistemology to contemporary issues*. New York: Blackwell.
El-Hani, C. N., Queiroz, J., & f. Stjernfelt. (2009). Firefly femmes fatales: A case study in the semiotics of deception. *Biosemiotics*, *1*, 33–55.
Evans, C. S. (2002). Cracking the code. Communication and cognition in birds. In M. Bekoff, C. Allen, & M. Burghardt (Eds.), *The cognitive animal. Empirical and theoretical perspectives on animal cognition* (pp. 315–322). Cambridge, MA: The MIT Press
Fetzer, J. K. (1990). *Artificial intelligence: Its scope and limits*. Dordrecht: Kluwer Academic Publisher.
Flach, P., & Kakas, A. (Eds.). (2000). *Abductive and inductive reasoning: Essays on their relation and integration*. Dordrecht: Kluwer Academic Publishers.
Frischen, A., Bayliss, A. P., & Tipper, S. P. (2007). Gaze cueing of attention. *Psychological Bulletin*, *133*(4), 694–724.
Gabbay, D. M., & Woods, J. (2005). *The reach of abduction*. Volume 2 of A practical logic of cognitive systems. Amsterdam: North-Holland.
Gibson, J. J. (1951). What is a form? *Psychological Review*, *58*, 403–413.
Gibson, J. J. (1979). *The ecological approach to visual perception*. Boston, MA: Houghton Mifflin.
Gibson, J. J. (1982). A preliminary description and classification of affordances. In E. S. Reed & R. Jones (Eds.), *Reasons for realism* (pp. 403–406). Hillsdale, NJ: Lawrence Erlbaum Associates Inc.
Girard, R. (1977). *Violence and the sacred* [1972]. Baltimore, MD: Johns Hopkins University Press.
Girard, R. (1986). *The Scapegoat* [1982]. Baltimore. MD: Johns Hopkins University Press.
Gooding, D. (1996). Creative rationality: Towards an abductive model of scientific change. *Philosophica*, *58*(2), 73–102.
Hanson, N. R. (1958). *Patterns of discovery: An inquiry into the conceptual foundations of science*. London: Cambridge University Press.
Harman, G. (1965). The inference to the best explanation. *Philosophical Review*, *74*, 88–95.
Harman, G. (1968). Enumerative induction as inference to the best explanation. *Journal of Philosophy*, *65*(18), 529–533.
Harman, G. (1973). *Thought*. Princeton, NJ: Princeton University Press.
Hebb, D. O. (1949). *The organization of behavior*. New York: John Wiley.
Josephson, J. R., & Josephson, S. G. (Eds.). (1994). *Abductive inference. Computation, philosophy, technology*. Cambridge: Cambridge University Press
Kuipers, T. A. F. (1999). Abduction aiming at empirical progress of even truth approximation leading to a challenge for computational modelling. *Foundations of Science*, *4*, 307–323.
Kuipers, T. (2000). *From instrumentalism to constructive realism. On some relations between confirmation, empirical progress and truth approximation*. Dordrecht: Kluwer Academic Publisher
Lipton, P. (2004). *Inference to the best explanation*. Routledge, London. Originally published in 1991. New Revised edition.
Loula, A., Gudwin, R., El-Hani., C. N., & Queiroz, J., (2010). Emergence of self-organized symbol-based communication in artificial creatures. *Cognitive Systems Research*, *2*, 131–147.
Magnani, L. (1988). Epistémologie de l'invention scientifique. *Communication and Cognition*, *21*, 273–291.
Magnani, L. (1992). Abductive reasoning: Philosophical and educational perspectives in medicine. In D. A. Evans & V. L. Patel (Eds.), *Advanced models of cognition for medical training and practice* (pp. 21–41). Berlin: Springer.
Magnani, L. (2001). *Abduction, reason, and science. Processes of discovery and explanation*. New York: Kluwer Academic/Plenum Publishers.
Magnani, L. (2009). *Abductive cognition. The eco-cognitive dimensions of hypothetical reasoning*. Heidelberg/Berlin: Springer.
Magnani, L. (2011). *Understanding violence. The intertwining of morality, religion, and violence: A philosophical stance*. Heidelberg/Berlin: Springer.

Peirce, C. S. (CP). *Collected papers of Charles Sanders Peirce* (Vols. 1–6). Cambridge, MA: Harvard University Press. In C. Hartshorne, & P. Weiss, (Eds.), vols. 7–8, Burks, A. W., ed., 1931–1958.
Peirce, C. S. (1966). *The Charles S. Peirce Papers: Manuscript Collection in the Houghton Library*. Worcester, MA: The University of Massachusetts Press. Annotated Catalogue of the Papers of Charles S. Peirce. Numbered according to Richard S. Robin. Available in the Peirce Microfilm edition. Pagination: CSP = Peirce / ISP = Institute for Studies in Pragmaticism.
Pinker, S. (1997). *How the mind works*. New York: W. W. Norton.
Pinker, S. (2003). Language as an adaptation to the cognitive niche. In M. H. Christiansen & S. Kirby (Eds.), *Language evolution* (pp. 16–37). Oxford: Oxford University Press.
Raftopoulos, A. (2009). *Cognition and perception. How do psychology and neural science inform philosophy*? Cambridge, MA: The MIT Press.
Shrager, J., & Langley, P. (Eds.). (1990). *Computational models of scientific discovery and theory formation*. San Mateo, CA: Morgan Kaufmann.
Simon, H. A. (1965). The logic of rational decision. *British Journal for the Philosophy of Science*, *16*, 169–186.
Simon, H. A. (1977). *Models of discovery and other topics in the methods of science*. Dordrecht: Reidel.
Taylor, K. (2009). *Cruelty, human evil and the human brain*. Oxford: Oxford University Press.
Thagard, P. (1987). The best explanation: Criteria for theory choice. *Journal of Philosophy*, *75*, 76–92.
Thagard, P. (1988). *Computational philosophy of science*. Cambridge, MA: The MIT Press.
Thom, R. (1972). *Stabilité structurelle et morphogénèse. Essai d'une théorie générale des modèles*. InterEditions, Paris. (D. H. Fowler. (1975). Trans. *Structural stability and morphogenesis: An outline of a general theory of models*. Reading, MA: W. A. Benjamin.
Thom, R. (1980). *Modèles mathématiques de la morphogenèse*. Paris: Christian Bourgois. (W. M. Brookes, & D. Rand. (1983). Trans. *Mathematical models of morphogenesis*. Chichester: Ellis Horwood.
Thom, R. (1988). *Esquisse d'une sémiophysique*. Paris: InterEditions. (V. Meyer. (1990). Trans. *Semio physics: A sketch*. Redwood City, CA: Addison Wesley).
Tooby, J., & DeVore, I. (1987). The reconstruction of hominid behavioral evolution through strategic modeling. In W. G. Kinzey (Ed.), *Primate models of hominid behavior* (pp. 183–237). Albany: Suny Press.
Waal, F. D., Wright, R., Korsgaard, C. M., Kitcher, P., & Singer, P. (Eds.). (2006). *Primates and philosophers. How morality evolved*. Princeton, NJ: Princeton University Press.
Walton, D. N. (2004). *Abductive reasoning*. Tuscaloosa, AL: The University of Alabama Press.
Žižek, S. (2009). *Violence* [2008]. London: Profile Books.

Part II
International Workshop Visual Abduction or Abductive Vision? KAIST (Korea Advanced Institute of Science and Technology)

Understanding Visual Abduction

The Need of the Eco-Cognitive Model

Lorenzo Magnani

> Abductive inference shades into perceptual judgment without any sharp line of demarcation between them.
> Charles Sanders Peirce, *Harvard Lectures on Pragmatism: Lecture VII*, 1903.

Abstract The status of abduction is still controversial. When dealing with abductive reasoning misinterpretations and equivocations are common. What did Peirce mean when he considered abduction both a kind of inference and a kind of instinct or when he considered perception a kind of abduction? Does abduction involve only the generation of hypotheses or their evaluation too? Are the criteria for the best explanation in abductive reasoning epistemic, or pragmatic, or both? Does abduction preserve ignorance or extend truth or both? To study some of these conundrums and to better understand the concept of *visual abduction*, I think that an interdisciplinary effort is needed, at the same time fecundated by a wide philosophical analysis. To this aim I will take advantage of some reflections upon Peirce's philosophy of abduction that I consider central to highlight the complexity of the concept, too often seen in the partial perspective of limited (even if tremendously epistemologically useful) formal and computational models. I will ponder over some seminal Peircean philosophical considerations concerning the entanglement of abduction, perception, and inference, which I consider are still important to current cognitive research. Peircean analysis helps us to better grasp how model-based, sentential, manipulative, and eco-cognitive aspects of *abduction*—I have introduced in my book *Abductive Cognition* (Magnani 2009)—have to be seen as intertwined, and indispensable for building an acceptable integrated model of visual abduction. Even if speculative, Peircean philosophical results on visual abduction certainly anticipate various tenets of recent cognitive research.

L. Magnani (✉)
Department of Humanities, Philosophy Section and Computational Philosophy Laboratory, University of Pavia, Pavia, Italy
e-mail: lmagnani@unipv.it

L. Magnani et al. (eds.), *Philosophy and Cognitive Science II*,
Studies in Applied Philosophy, Epistemology and Rational Ethics 20,
DOI 10.1007/978-3-319-18479-1_7

1 Perception Versus Inference in Abductive Cognition

We should remember, as Peirce noted, that abduction plays a role even in relatively simple visual phenomena. *Visual abduction*,[1] a special form of non verbal abduction—a kind of model-based cognition—occurs when hypotheses are instantly derived from a stored series of previous similar experiences. It covers a mental procedure that falls into the category called "perception". Peirce considers *perception* a fast and uncontrolled knowledge-production process. Perception is a kind of vehicle for the instantaneous retrieval of knowledge that was previously assembled in our mind through inferential processes. Keeping in mind Peirce's famous syllogistic framework for abduction (as a form of fallacy of the affirming the consequent) we can say that, in the case of perception we face with a situation in which: "[...] a fully accepted, simple, and interesting inference tends to obliterate all recognition of the uninteresting and complex premises from which it was derived" (Peirce 1931–1958, 7.37). We can add that many visual stimuli—that can be considered the "premises" of the involved abduction—are ambiguous, yet people are adept at imposing order on them: "We readily form such hypotheses as that an obscurely seen face belongs to a friend of ours, because we can thereby explain what has been observed" (Thagard 1988, p. 53). This kind of image-based hypothesis formation can be considered as a form of *visual abduction*. Hence, perception is abductive in itself: "Abductive inference shades into perceptual judgment without any sharp line of demarcation between them" (Peirce 1992–1998, p. 224). Visual abduction plays an important cognitive role in both everyday reasoning and science, where it is well known it can provide epistemically substantial shortcuts to dramatic new discoveries.

If perceptions are abductions they are basically withdrawable, just like the scientific hypotheses abductively found. In this perspective perceptions can be seen as "hypotheses" about data we can accept (usually this happens spontaneously) or carefully evaluate. Moreover, the fact they can be considered, as we will see in the Sect. 2.1, *inferences*, in the Peircean sense, and so withdrawable, does not mean they are controlled (deliberate), like in the case of explicit inferences, for example in logic and other types of rational or fully conscious human reasoning. Perception involves semiosis and is abductive, and it is able to correct itself when it falls into error, and consequently it can be censured. However, we have to carefully analyze the proper character of this kind of controllability, following Peirce's considerations on the so-called "perceptual judgment" ("The seven systems of metaphysics", 1903):

> Where then in the process of cognition does the possibility of controlling it begin? Certainly not before the *percept* is formed. Even after the percept is formed there is an operation, which seems to me to be quite uncontrollable. It is that of judging what it is that the person perceives. A judgment is an act of formation of a mental proposition combined with an adoption of it or act of assent to it. A percept on the other hand is an image or moving picture or other exhibition. [...] I do not see that it is possible to exercise any control over that operation or to subject it to criticism. If we can criticize it at all, as far as I can see, that criticism would be limited to performing it again and seeing whether with closer attention we get

[1] I have introduced the basic aspects of visual abduction in Magnani et al. (1994) and Magnani (1996).

> the same result. But when we so perform it again, paying now closer attention, the percept is presumably not such it was before. I do not see what other means we have of knowing whether it is the same as it was before or not, except by comparing the former perceptual judgment to the later one. I would utterly distrust any other method of ascertaining what the character of the percept was. Consequently, until I am better advised, I shall consider the *perceptual judgment* to be utterly beyond control (Peirce 1992–1998, II, p. 191).

In summary, judgments in perception are fallible but indubitable abductions—we are not in any condition to psychologically conceive that they are false, as they are unconscious habits of inference.

Nevertheless, percept and perceptual judgment are not unrelated to abduction because they are not entirely free

> [...] from any character that is proper to interpretations [...]. The fact is that it is not necessary to go beyond ordinary observations of common life to find a variety of widely different ways in which perception is interpretative. The whole series of hypnotic phenomena, of which so many fall within the realm of ordinary everyday observation, – such as waking up at the hour we wish to wake much nearer than our waking selves could guess it, – involve the fact that we perceive what we are adjusted for interpreting though it be far less perceptible that any express effort could enable us to perceive [...]. It is a marvel to me that the clock in my study strikes every half an hour in the most audible manner, and yet I never hear it [...]. Some politicians think it is a clever thing to convey an idea which they carefully abstain from stating in words. The result is that a reporter is ready to swear quite sincerely that a politician said something to him which the politician was most careful not to say. It is plainly nothing but the extremest case of Abductive Judgment (Peirce 1992–1998, II, p. 229).

1.1 Perceptions, Iconic Cognition, and Model-Based Abduction

The fact that perception functions as a kind of "abstractive observation" (Peirce 1992–1998 II, p. 206), so that "perceptual judgments contain general elements" (Peirce 1992–1998, II, p. 227) relates it to the expressive power of icons. It is analogous to what is occurring in mathematics when the reasoner "sees"—through the manipulations and constructions on an external single diagram (icon)—that some properties are not merely single but of a general nature: perception functions as "an abstractive observation". Indeed Peirce was clearly aware, speaking of the model-based aspects of deductive reasoning, that there is an "experimenting upon this image [for example the external model/diagram] in the imagination", where the idea that human imagination is always favored by a kind of prosthesis, the external model as an "external imagination", is pretty clear, even in the case of classical geometrical deduction: "[...] namely, deduction consists in constructing an icon or diagram the relations of whose parts shall present a complete analogy with those of the parts of the object of reasoning, of experimenting upon this image in the imagination and of observing the result so as to discover unnoticed and hidden relations among the parts" (Peirce 1931–1958, 3.363). Peirce eloquently concludes that it is "[...] a very extraordinary nature of Diagrams that they show—as literally as Percept shows the Perceptual Judgment to be true,—that a consequence does follow, and more marvelously yet, that it would follow under all varieties of circumstances accompanying the premises" (Peirce 1976, pp. 317–318).[2]

These Peircean considerations also echo the Kantian ones concerning geometry. Immanuel Kant was clearly aware of the interplay between internal and external models—exemplified in the case of a formal science like mathematics—as an example of genuine knowledge production (and, occasionally, of discovery). In his transcendental terms, Kant says that in geometrical construction "[...] I must not restrict my attention to what I am actually thinking in my concept of a triangle (this is nothing more than the mere definition); I must pass beyond it to properties which are not contained in this concept, but yet belong to it" (Kant 1929, A718-B746, p. 580). Hence, for Kant models in science (in this case, of geometry) are first of all *constructions* that go beyond what the researcher simply "thinks", and exploit "external" representations: to solve the classical geometrical problem of the sum of the internal angles of a triangle, the agent for example "[...] begins by *constructing* a triangle. Since he knows that the sum of two right angles is exactly equal to the sum of all the adjacent angles which can be constructed from a single point on a straight line, he prolongs one side of his triangle and obtains two adjacent angles, which together are equal to two right angles. He then divides the external angle by drawing a line parallel to the opposite side of the triangle, and observes that he has thus obtained an external adjacent angle which is equal to an internal angle—and so on. In this fashion, through a chain of inferences guided throughout by intuition, he arrives at a fully evident and universally valid solution of the problem" [(Kant 1929, A716-B744, pp. 578–579), emphasis added].

2 Iconicity Hybridates Logicality: Inference in a Semiotic Perspective

2.1 Is Perception an Inference?

Let us consider some further basic philosophical aspects related to the problem of perception introduced by Peirce. In the following passage, which Peirce decided to skip in his last of the seven Harvard Lectures (14 May 1903), perception is clearly considered a kind of abduction: "A mass of facts is before us. We go through them. We examine them. We find them a confused snarl, an impenetrable jungle. We are unable to hold them in our minds. [...] But suddenly, while we are poring over our digest of the facts and are endeavoring to set them into order, it occurs to us that if we were to assume something to be true that we do not know to be true, these facts

[2]Cf. Turrisi (1990). Other considerations on abduction and perception are given in Tiercelin (2005).

would arrange themselves luminously. That is *abduction* [...]".[3] This passage seems to classify abduction as emerging in "perceiving" facts and experiences, and not only in the conclusions of an "inference" (Hoffmann 1999, pp. 279–280), intended in the classical sense, as expressed by symbols carrying propositional content.

Let us reiterate the following passage, already quoted above in Sect. 1; if we say that by perception, knowledge constructions are so instantly reorganized that they become habitual and diffuse and do not need any further testing: "[...] a fully accepted, simple, and interesting inference tends to obliterate all recognition of the uninteresting and complex premises from which it was derived" (Peirce 1931–1958, 7.37). I also noted: many visual stimuli—that can be considered the "premises" of the involved abduction—are ambiguous, yet people are adept at imposing order on them. Woods comments, suspecting the limitations of the so-called (formal) GW-model of abduction[4]: "Perceptual abduction is interesting in a number of ways. As a fast and uncontrolled knowledge production, it operates for the most part automatically and out of sight, so to speak. If true, this puts a good deal of pressure on any suggestion that the GW-schema might be canonical for abduction. In its present formulation, what the schema schematizes is sentential abduction, as Magnani calls it; that is, abduction rendered by symbols carrying propositional content" (Woods 2011, p. 242).

The semio-philosophical literature on abduction has afforded the conciliation between the sentential and the perceptual aspects of abduction trying to subordinate the second to the first: at a certain level of abstraction, visual stimuli, for example, can be viewed as premisses, and the outputs of perceptual processing—our knowledge that an obscurely seen face belongs to a friend of ours—in turn be likened to a conjecture derived from the fact or apparent fact that the best causal account of the presence of those stimuli is the presence of our friend. Woods concludes: "In fact, a rather common answer is that what we are told when it is claimed that at a certain level of abstraction perception is hypothesis-drawing from premisses is that perception is *tacit* hypothesis-drawing from premisses; that the processes that take visual stimuli to a knowledge of birds is abductively inferential in character, but unconsciously and non-symbolically so-*implicitly*."

Certainly the processes that generate perceptual knowledge can be modeled as abduction adopting the subordination of the perceptual side to the sentential one, but I would prefer a coexistence between model-based and sentential aspects, rather than the above conciliation. I endorse a compromise, which I think can increase the intelligibility of abduction as a wide way of inferring hypotheses: given that the concept of abduction is not at all exhausted by the formal models of it, the concept of abduction can be better understood in the light of a composite eco-cognitive view,[5] exactly following the spirit of Peirce's philosophy. If we rigidly separate the two aspects, the inferential one (using the adjective "inferential" just to refer to logical

[3]Cf. "Pragmatism as the logic of abduction", in Peirce (1992–1998, pp. 227–241), the quotation is from footnote 12, pp. 531–532.

[4]See below the Appendix: GW and AKM schemas of abduction.

[5]On my eco-cognitive model (EC-model) of abduction cf. Magnani (2013).

and propositional accounts), and the perceptual (as referred to the model-based, or more in general, non sentential semiotic accounts), we rejoin the intellectual conundrum already present in the literature on abduction, caused by the suspected manifest inconsistency of the two views. In this perspective perceptual and inferential views are contrasted and a kind of inconsistency arises, as many researchers contend.

Indeed, it is well-known that in Peirce the inferential side of abduction is initially expressed and denoted by the logico-syllogistic framework. We have just illustrated that, following this point of view the genesis of an—abductive—perceptual judgment would have to be located, following some interpreters, at the level of the premises of the famous Peircean syllogistic schema, that depicts abduction as the fallacy of the affirming the consequent. Moreover, it would be at the level of this perceptual side, and *not* at the level of the logico-sentential one that the proper creative virtues of abduction would be disclosed. The explaining solution would emerge in perceiving facts and experience and not in the conclusion of the logical inference ["the initial conceiving of a novel hypothesis is not the product of an inferential transition" (Kapitan 1997, p. 2)].[6]

As I have anticipated, I think that the two—often considered contrasting—views more simply and coherently can coexist, beyond Peirce,[7] but also in the perspective of the orthodoxy of Peircean texts: the prevailing Peircean *semiotic* conception of inference as a form of sign activity, where the word sign includes "feeling, image, conception, and other representation" offers the solution to this potential conflict. For example, Anderson (1987, p. 45) maintains that "Peirce quite explicitly states that abduction is both an insight and an inference. This is a fact to be explained, not to be explained away". Anderson nicely solves this problem by referring to Peirce's theory of the three fundamental categories, Firstness, Secondness, and Thirdness: abduction, as a form of reasoning is essentially a third, but it also occurs at the level of Firstness "as a sensuous form of reasoning" (p. 56 ff.).

In my perspective the meaning of the word inference is not exhausted by its "logical" aspects but is referred to the effect of various sensorial activities. One more reason that supports my contention is that for Peirce the sentential aspects of symbolic disciplines like logic or algebra coexist with model-based features—iconic. Sentential features like symbols and conventional rules are intertwined with the spatial configuration; in Peirce's terms we have already quoted above:

> The truth, however, appears to be that all deductive reasoning, even simple syllogism, involves an element of observation; namely, deduction consists in constructing an icon or diagram the relations of whose parts shall present a complete analogy with those of the parts of the object of reasoning, of experimenting upon this image in the imagination and of observing the result so as to discover unnoticed and hidden relations among the parts (Peirce 1931–1958, 3.363).

[6]It has to be said that some authors [for example Hoffmann (1999, p. 280)] contend that, in order to explain abduction as the process of forming an explanatory hypothesis within Peirce's concept of "logic", it is necessary to see both sides as coming together.

[7]It is well-known that in later writings Peirce seems more inclined to see abduction as both insight and inference.

2.2 The Compound Conventional Sign

In another passage, which refers to the "conventional" character of algebraic formulas as icons, the hybridity between sentential and model-based aspects is even clearer and takes advantage of the introduction of the idea of the "compound conventional sign"[8]:

> Particularly deserving of notice are icons in which the likeness is aided by conventional rules. Thus, an algebraic formula is an icon, rendered such by the rules of commutation, association, and distribution of the symbols; that it might as well, or better, be regarded as a compound conventional sign (Peirce 1966, p. 787 and pp. 26–28 CSP).

It seems for Peirce that iconicity of reasoning, and consequently of abduction are fundamental, like it is clearly stressed in the following further passage: "I said, Abduction, or the suggestion of an explanatory theory, is inference through an Icon" (Peirce 1986, p. 276). Moreover, induction and deduction are inferences "through an Index" and "through a Symbol" (*ibid.*).

To summarize, it would seem that there is not an inferential aspect of abduction, characterized by the syllogistic model, *separated* from (or contrasted with) the perceptual one, which would be "creative" instead, as many authors contend.[9] I consider the two aspects consistent, and both are perfectly understandable in the framework of Peircean philosophy and semiotics.[10]

A further evidence of the fact that the two aspects of abduction are intertwined derives from the study of children's early word acquisition (Roberts 2004). Children form knowledge and expectations about the symbolic functioning of a particular word in routine events where model-based perceptual and manipulative aspects of reasoning are predominant and furnish suitable constraints: they generate abductions that help to acquire the content-related symbolic functioning, going beyond what was already experienced. These abduced hypotheses are "practical", about knowing how to use a word to direct attention in a certain way. These hypotheses need not be verbalized by the children, who only later on acquire a more theoretical status through a systematization of their knowledge. It is at this level that they are expressed verbally and concern causal frameworks rather than specific causal mechanisms—for instance of natural kind terms.

To further deepen the particular "inferential" status of abduction we have illustrated above, further problems regarding the relationship between sentential and model-based aspects of abduction have to be analyzed.

---

[8]Stjernfelt (2007) provides a full analysis of the role of icons and diagrams in Peircean philosophical and semiotic approach, also taking into account the Husserlian tradition of phenomenology.

[9]For example Kapitan (1997), Hoffmann (1999).

[10]On the contrary, some authors [for example Hoffmann (1999, 2004), Paavola (2004)], find a central paradox in what (Frankfurt 1958, p. 594) clearly synthesized by saying "[...] that Peirce holds both that hypotheses are the products of a wonderful imaginative faculty in man and that they are product of a certain sort of logical inference". Furthermore, some commentators seem to maintain that "creative" aspects of abduction would exclusively belong to the perceptual side, as I have already noted above.

2.3 Syllogism Versus Perception?

The following is a frequently quoted passage by Peirce on perception and abduction related to the other passage on "perceptual judgment" that I reported above at the beginning of Sect. 1:

> Looking out of my window this lovely spring morning I see an azalea in full bloom. No no! I do not see that; though that is the only way I can describe what I see. *That* is a proposition, a sentence, a fact; but what I perceive is not proposition, sentence, fact, but only an image, which I make intelligible in part by means of a statement of fact. This statement is abstract; but what I see is concrete. I perform an abduction when I so much as express in a sentence anything I see. The truth is that the whole fabric of our knowledge is one matted felt of pure hypothesis confirmed and refined by induction. Not the smallest advance can be made in knowledge beyond the stage of vacant staring, without making an abduction in every step.[11]

The classical interpretation of this passage stresses the existence of a vicious circle (Hoffmann 1999, p. 283). On the one hand, we learn that the creativity of abduction is based on the genesis of perceptual judgments. On the other hand, it is now said that any perceptual judgment is in itself the result of an abduction. Or, as Peirce says, "[...] our first premises, the perceptual judgments, are to be regarded as an extreme case of abductive inference, from which they differ in being absolutely beyond criticism" (Peirce 1931–1958, 5.181).

Surely it can be maintained that for Peirce perception on the whole is more precisely the act of subsuming sense data or "percepts" under concepts or ideas to give rise to perceptual judgments: we have just said in the previous subsection that he in turn analyzed this act of subsuming as an abductive inference depicted in syllogistic terms

> (P1) A well-recognized kind of object, *M*, has for its ordinary predicates P[1], P[2], P[3], etc.
>
> (P2) The suggesting object, *S*, has these predicates P[1], P[2], P[3], etc.
>
> (C) Hence, *S* is of the kind *M* (Peirce 1931–1958, 8.64).

In this abductive inference—which actually is merely "selective"—the creative act "would" take place in the second premise: if we distinguish in abduction an inferential part and a perceptual one—cf. above—(that is the genesis of a perceptual judgment), and if we understand according to Peirce the arising of a perceptual judgment for itself as an abductive inference, then in explaining the possibility of abduction we get an infinite regress. Fortunately, Peirce notes, the "process of forming the perceptual judgment" is "sub-conscious and so not amenable to logical criticism", hence, it is not discrete like sentential inferences, but a "continuous process":

> On its side, the perceptive judgment is the result of a process, although of a process not sufficiently conscious to be controlled, or, to state it more truly, not controllable and therefore not fully conscious. If we were to subject this subconscious process to logical analysis, we

[11] Cf. the article "The proper treatment of hypotheses: a preliminary chapter, toward an examination of Hume's argument against miracles, in its logic and in its history" [1901] [in Peirce (1966, p. 692)].

> should find that it terminated in what that analysis would represent as an abductive inference, resting on the result of a similar process which a similar logical analysis would represent to be terminated by a similar abductive inference and so on *ad infinitum*. This analysis would be precisely analogous to which the sophism of Achilles and the Tortoise applied to the chase of the Tortoise by Achilles, and it would fail to represent the real process for the same reason. Namely, just as Achilles does not have to make the series of distinct endeavors which he is represented as making, so this process of forming the perceptual judgment, because it is sub-conscious and so not amenable to logical criticism, does not have to make separate acts of inference, but performs its act in one continuous process (Peirce 1931–1958, 5.181).

This recursiveness, and the related vicious circle, even if stressed by many commentators, do not seem to me really important. I think we can give a simpler explanation of this conflict between the inferential and perceptual side of abduction by recalling once again the Peircean *semiotic* conception of inference as a form of sign activity, where the word sign includes "feeling, image, conception, and other representation".

3 The Need of an Eco-Cognitive Model [EC-Model] of Abduction

3.1 Perception and Abduction as an Inference to the Best Explanation

In the standard accounts of abductive reasoning, abduction as an inference to the best explanation also involves what Peirce called the inductive/evaluative phase. It is clear that viewing perception as an abduction hardly fits this standard view. No need of empirical evaluation in perception, and consequently it cannot be said that testability is intrinsic to abduction, such as Peirce himself seems to contend in some passages of his writings. In perception the "best abductive choice" is immediately reached—in an uncontrolled way—without the help of an experimental trial (which fundamentally characterizes the received view of abduction in terms of the so-called "inference to the best explanation"). Not only, we have to strongly note that the generation process alone can suffice: in perception the hypothesis generated is immediate and unique.

At the center of my perspective on abductive cognition is the emphasis on the "practical agent", of the individual agent operating "on the ground", that is, in the circumstances of real life. In all its contexts, from the most abstractly logical and mathematical to the most roughly empirical, I always emphasize the cognitive nature of abduction. Reasoning is something performed by cognitive systems. At a certain level of abstraction and as a first approximation, a cognitive system is a triple (A, T, R), in which A is an *agent*, T is a *cognitive target* of the agent, and R relates to the *cognitive resources* on which the agent can count in the course of trying to meet the target-information, time and computational capacity, to name the three most important. My agents are also *embodied distributed cognitive systems*: cognition is embodied and the interactions between brains, bodies, and external environment are its central aspects. Cognition is occurring taking advantage of a constant exchange of information in a complex distributed system that crosses the boundary between humans, artifacts, and the surrounding environment, where also instinctual and unconscious abilities play an important role. This interplay is especially manifest and clear in various aspects of abductive cognition.[12]

It is in this perspective that we can appropriately consider perceptual abduction—as I have already said—as a fast and uncontrolled knowledge production, that operates for the most part automatically and out of sight, so to speak. This means that—at least in this light—GW-schema is not canonical for abduction, as I have already pointed out. The schema illustrates what I call "sentential abduction" (Magnani 2009, Chap. 1), that is, abduction rendered by symbols carrying propositional content. It is hard to encompass in this model cases of abductive cognition such as perception or the generation of models in scientific discovery.[13] My perspective adopts the wide Peircean philosophical framework, which approaches "inference" *semiotically* (and not simply "*logically*"). It is clear that this semiotic view is considerably compatible with my perspective on cognitive systems as embodied and distributed systems: the GW-schema is instead only devoted to illustrate, even if in a very efficacious way, a subset of the cognitive systems abductive activities, the ones that are performed taking advantage of explicit propositional contents. Woods seems to share this conclusion: "[…] the GW-model helps get us started in thinking about abduction, but it is nowhere close, at any level of abstraction, to running the whole show. It does a good job in modelling the ignorance-preserving character of abduction; but, since it leaves the S_i of the schema's clause (T) unspecified, it makes little contribution to the fill-up problem" (Woods 2011, p. 244).

In the perspective of my eco-cognitive model (EC-model) the cutdown problem (that is the problem of specifying the conditions for *thinking up* possible candidates for selection) and the fill-up one (that is the problem of finding criteria for hypothesis *selection*) in abductive cognition appear to be spectacularly *contextual*.[14] I lack the space to give this issue appropriate explanation but it suffices for the purpose of this study to remember that, for example, one thing is to abduce a model or a concept at the various levels of scientific cognitive activities, where the aim of reaching rational knowledge dominates, another thing is to abduce a hypothesis in literature (a fictional character for example), or in moral reasoning (the adoption/acceptation of a hypothetical judgment as a trigger for moral actions). The case of perception is extreme, because in this case abduction is in itself unconscious and automatic—and immediately "accepted", so to speak—and of course evidentially inert (in the sense that there is no need of the empirical evaluation, instead mandatory in the case of the

---

[12]It is interesting to note that recent research on Model Checking in the area of AST (Automated Software Testing) takes advantage of this eco-cognitive perspective, involving the manipulative character of model-based abduction in the practice of adapting, abstracting, and refining models that do not provide successful predictions, cf. Angius (2013).

[13]On the knowledge enhancing role of abduction in guessing models in science cf. Magnani (2013).

[14]Some acknowledgment of the general contextual character of these kinds of criteria, and a good illustration of the role of coherence, unification, explanatory depth, simplicity, and empirical adequacy in the current literature on *scientific* abductive best explanation, is given in Mackonis (2013).

appropriate activation of a hypothesis in the more composite abductive processes in empirical science). To conclude, the proper experimental test involved in the Peircean evaluation—inductive—phase, which for many researchers would reflect in the most acceptable way the idea of abduction as inference to the best explanation—and so carrier of new reliable knowledge—just constitutes a *special* subclass of the multiple possible modes of adoption/acceptation of an abductive hypothesis.

The backbone of my approach can be found in the manifesto of my eco-cognitive model (EC-model) of abduction in Magnani (2009). It might seem awkward to speak of "abduction of a hypothesis in literature," but one of the fascinating aspects of abduction is that not only it can warrant for scientific discovery, but for other kinds of creativity as well. We must not necessarily see abduction as a *problem solving device* that sets off in response to a cognitive irritation/doubt: conversely, it could be supposed that aesthetic abductions (referring to creativity in art, literature, music, etc.) arise in response to some kind of aesthetic irritation that the author (sometimes a *genius*) perceives in herself or in the public. Furthermore, not only aesthetic abductions are usually free from empirical constraints in order to become the "best" choice: as I am showing throughout this paper, many forms of abductive hypotheses in traditionally-perceived-as-rational domains (such as the setting of initial conditions, or axioms, in physics or mathematics) are relatively free from the need of an empirical assessment. The same could be said of moral judgement: they are eco-cognitive abductions, inferred upon a range of internal and external cues and, as soon as the judgment hypothesis has been abduced, it immediately becomes prescriptive and "true," informing the agent's behavior as such. Assessing that there is a common ground in all of these works of what could be broadly defined as "creativity" does not imply that all of these forms of creativity are the same, contrarily it should spark the need for firm and sensible categorization: otherwise it would be like saying that to construct a doll, a machine-gun and a nuclear reactor are all the same thing because we use our hands in order to do so!

4 Explicit, Uncontrolled, and Unconscious Inferences in Multimodal Abduction

As I have maintained in the previous sections, I think that two contrasting views of abduction such as inferential and model-based (like in the case of perception) can coherently coexist: I have already contended that the prevailing Peircean *semiotic* conception of inference as a form of sign activity offers the solution to the conflict. The EC-model I have outlined in the previous section reinforces this view in a broad cognitive framework. We also said that for Peirce the sentential aspects of logic, even if central, coexist with model-based features—iconic. Abduction can be performed by words, symbols, and logical inferences, but also by internal processes that treat external sensuous input/signs through merely unconscious mechanisms which give rise to abductive actions and reactions, like in the case of the well-known instinctive reactions of the humble Peircean chicken [cf. Magnani (2009, Chap. 5)] or of human emotions and other various implicit ways of thinking. In these last cases sentential aspects do not play any role (or a dominant role).

We can say, following Thagard (2005, 2007) that abduction is fundamentally performed in a *multimodal* way: for example, we consciously perform a perceptual judgment about the azalea, and in this case also concepts, ideas and statements certainly play a central abductive role, but—Peirce says, they are only *part* of the whole process: "what I perceived is not proposition, sentence, fact, but only image, which I made intelligible in part by means of a statement of fact".[15] It is in this way that perceptions acquire "meanings": they nevertheless remain "hypotheses" about data we can accept (usually this happens spontaneously) or carefully submit to criticism. It is in this sense that the visual model of perception does not work in isolation from other modes of perception or from other persons or sources of experience (Gooding 1996). As I have already illustrated in the first section above perceptions are withdrawable "inferences", even if not controlled (deliberate), like we control explicit inferences for example in logic and other types of more or less rational human "reasoning" and argumentation.

Being creative is not a peculiarity of perceptual/visual abduction, like—as I have already said—some commentators seem to maintain (Kapitan 1997). Moreover, perception and cognition alike are inherently inferential. If awareness, whether propositional or perceptual, is semiotic, then all awareness involves the interpretation of signs, and all such interpretation is inferential: semiosis not only involves the interpretation of *linguistic* signs, but also the interpretation of *non-linguistic* signs. Abduction of course embraces much of these semiotic performances.

In sum, from a naturalistic perspective both linguistic and non linguistic signs

1. have an internal semiotic life, as particular configurations of neural networks and chemical distributions (and in terms of their transformations) at the level of human brains, and as somatic expressions,
2. but can also be delegated to many external objects and devices, for example written texts, diagrams, artifacts, etc.

In this "distributed" framework those central forms of abductive cognition that occur in a hybrid way, that is in the interplay between internal and external signs, are of special interest: abduction can be properly seen only in an eco-cognitive framework.

5 Perception Is Semi-Encapsulated

Recent cognitive studies on perception seem to confirm Peirce's philosophical speculations. Through an interdisciplinary approach and suitable experimentation some cognitive scientists [cf. for example Raftopoulos (2001a, b)] have acknowledged the

[15] Cf. "The proper treatment of hypotheses: a preliminary chapter, toward an examination of Hume's argument against miracles, in its logic and in its history" (1901) [in Peirce (1966, p. 692)].

fact that, in humans, perception (at least in the visual case) is not strictly modular, as Fodor (1984) argued, that is, it is not encapsulated, hardwired, domain-specific, and mandatory.[16] Neither is it wholly abductively "penetrable" by higher cognitive states (like desires, beliefs, expectations, etc.), by means of top-down pathways in the brain and by changes in its wiring through perceptual learning, as stressed by Churchland (1988). It is important to consider the three following levels: visual sensation (bodily processes that lead to the formation of retinal image which are still useless—so to speak—from the high-level cognitive perspective), perception (sensation transformed along the visual neural pathways in a structured representation), and observation, which consists in all subsequent visual processes that fall within model-based/propositional cognition. These processes "[…] include both post-sensory/semantic interface at which the object recognition units intervene as well as purely semantic processes that lead to the identification of the array—high level vision" (Raftopoulos 2001b, p. 189).[17]

On the basis of this distinction it seems plausible—as Fodor contends—to think there is a substantial amount of information in perception which is theory-neutral. However, also a certain degree of theory-ladenness is justifiable, which can be seen at work for instance in the case of so-called "perceptual learning". However, this fact does not jeopardize the assumption concerning the basic cognitive impenetrability of perception: in sum, perception is informationally "semi-encaspulated", and also semi-hardwired, but, despite its bottom-down character, it is not insulated—so to speak—from "knowledge". For example, it results from experimentation that illusion is a product of learning from experience, but this does not regard penetrability of perception because these experience-driven changes do not affect a basic core of perception.[18]

Higher cognitive states affect the product of visual modules only after the visual modules "[…] have produced their product, by selecting, acting like filters, which output will be accepted for further processing" (Raftopoulos 2001a, p. 434), for instance by selecting through attention, imagery, and semantic processing, which aspects of the retinal input are relevant, activating the appropriate neurons. I have tried to show in this article that I consider these processes essentially abductive, as is also clearly stressed by Shanahan (2005), who provides an account of robotic perception from the perspective of a sensory fusion in a unified framework: he describes problems and processes like the incompleteness and uncertainty of basic sensations, top-down information flow and top-down expectation, active perception and attention.[19]

[16]Challenges to the modularity hypothesis are illustrated in Marcus (2006).

[17]A full treatment of the problem of perception both from a psychological and neural perspective is available in the recent (Raftopoulos 2009). A recent rich volume that shows the semi-encapsulated character of perception as illustrated by recent cognitive science results is Albertazzi et al. (2011).

[18] Evidence on the theory-ladenness of visual perception derived from case-studies in the history of science is illustrated in Brewer and Lambert (2001).

[19]Cohn et al. (2002) propose a cognitive vision system based on abduction and qualitative spatio-temporal representations capable of interpreting the high level semantics of dynamic scenes. Banerjee (2006) presents a computational system able to manage events that are characterized by a large number of individual moving elements, either in pursuit of a goal in groups (as in military

It is in this sense that a certain amount of *plasticity* in vision does not imply the full penetrability of perception. As I have already noted, this result does not have to be considered equivalent to the claim that perception is, so to speak, not theory-laden. It has to be acknowledged that even basic perceptual computations obey high-level constraints acting at the brain level, which incorporate implicit and more or less model-based assumptions about the world, coordinated with motor systems. At this level, they lack a semantic content, so as they are not learnt, because they are shared by all, and fundamentally hardwired.

Human auditory perception should also be considered semi-encapsulated (de Cheveigné 2006). The human auditory system resembles that of other vertebrates, such as mammals, birds, reptiles, amphibians or fish, and it can be thought to derive from simple systems that were originally strictly intertwined with motor systems and thus linked to the sense of space.[20]

Hearing, which works in "dark and cluttered" (de Cheveigné 2006, p. 253) environments, is complementary to other senses, and has both neural bottom-up and top-down characters. The top-down process takes advantage of descending pathways that send *active* information out from a central point and play a part in selectively "listening" to the environment, involving relevant motor aspects (indeed action is fundamental to calibrating perception). The role of hearing in the perception of space is central, complementing multichannel visual information with samples of the acoustic field picked up by the ears: cues to location of source by means of interaural intensity, difference and distance according to cues like loudness are two clear examples of the abductive *inferential* processes performed by hearing that provide substantial models of the scene facing the agent. The whole process is abductive in so far as it provides selections of cues, aggregation of acoustic fragments according to source and an overall hypothetical meaningful explanation of acoustic scenes, that are normally very complex from the point of view of the plurality of acoustic sources. The auditory system of vertebrates which decouples perception from action (motor systems)—still at work together in acoustically rudimentary organisms—enhances economy, speed, and efficacy of the cognitive system by exploiting abstract models of the environment and motor plans.

(Footnote 19 continued)
operations), or subject to underlying physical forces that group elements with similar motion (as in weather phenomena). Visualizing and reasoning about happenings in such domains are treated through a multilayered abductive inference framework where hypotheses largely flow upwards from raw data to a diagram, but there is also a top-down control that asks lower levels to supply alternatives if the higher level hypotheses are not deemed sufficiently coherent.

[20]The example of a simple hypothetical organism equipped with two fins and two eyes (Szentagothai and Arbib 1975) can explain this link between perception and action in the case of vision: "The right eye was connected to the left fin by a neuron, and the left eye to the right fin. When a prey appears within the field of the right eye, a command is sent to the left fin to instruct it to move. The organism then turns towards the prey, and this orientation is maintained by bilateral activation until the prey is reached. *Perception* in this primitive organism is not distinct from action" (de Cheveigné 2006, pp. 253–254).

The digression above about abductive cognition as perception is certainly endowed with an eco-cognitive character, individual human agents, senses, environment, and cognition are all involved. How can we better grasp the complete significance of what I called (cf. Sect. 3) eco-cognitive model of abduction? As I have illustrated we need refer to the cognitive science tradition of embodied and distributed cognitive systems: at the of center of my perspective on abductive cognition is the emphasis on the "practical agent", on the individual agent operating "on the ground", that is, in the circumstances of real life.

Hence, to better explain my eco-cognitive approach to abduction let me finally describe some basic concepts regarding distributed and embodied cognitive systems. Early work in distributed cognition was informed by the basic idea that cognition is a socially distributed phenomenon, one that is situated in actual practices. The theory contends that cognitive processes generally are best understood as situated in and distributed across concrete socio-artifactual contexts. The received theories in cognitive science emphasize an internalism that relegates to a lower edge the role of external representations (and so of the correspondent cognitive delegations to various areas of the external environment) and problem solving in collaborative contexts.

The new theoretical view criticizes the traditional accounts of cognition, emphasizing instead the role of concrete social and artifactual contexts. At the same time a kind of ecological perspective stresses the role played by the agent-environment interaction. For example, in current collaborative work environments, we find humans and artifactual technologies together preserving and manipulating representational states, very often to the aim of solving problems. The theory of distributed cognition is motivated by the idea that such complex systems perform authentic cognitive processes and that the cognitive features and properties of these kinds of socially, materially and temporally distributed systems differ from those of the agents that act in them.

The theory of distributed cognition was proposed by Edwin Hutchins to present a new analysis of problem solving processes in real work settings, and to supply a new framework for cognitive science generally. In his seminal study Hutchins (1995) describes how agents use tools and instruments (and so external cognitive representations) to produce, create, manipulate, and maintain representational states. Hutchins contends that the cognitive properties of the distributed cognitive system depend on the physical and "material" properties of the external representational media in which they are applied. The theory of distributed cognition does not destroy the concept of individual cognition, even if eco-cognitive aspects are emphasized. The aim is the analysis of cognition as distributed across people and artifacts, and of its strict reference to the interplay of how both internal and external representations.[21] It is in the theoretical framework I have just described that my eco-cognitive approach to abduction has to be seen.

---

[21] On this interplay and on the role of external representations as material anchors for conceptual blends see also the more recent (Hutchins 2005).

6 Conclusion

I this article I have illustrated that, to understand visual abduction, an "archeological"—and at the same time interdisciplinary—effort is mandatory, which takes advantage of both the critical revision of philosophical classical speculations and recent epistemological and cognitive results. To this aim I have analyzed some "canonic" aspects of Peirce's philosophy resorting to the description of the role of abduction in inferences, in perception, and in diagrams and icons, and I have quoted the case of abduction as instinct-based. I have further intertwined these traditional issues to the recent analysis of creative, selective, model-based, multi-modal, and manipulative abduction, adopting an extended and rich eco-cognitive perspective (EC-model of abduction). Following this intellectual route, *understanding visual abduction* also becomes a way of better recognizing the limitations of formal and computational models, otherwise so useful to focus on other relevant aspects of abductive reasoning, such as ignorance-preservation, relevance and plausibility criteria, and the problem of the inference to the best explanation. Peircean analysis helps us to better grasp how sentential, model-based (and so "visual"), and manipulative aspects of abduction have to be seen as intertwined, and indispensable for building a satisfactory and unified model of abduction.

Appendix: GW and AKM Schemas of Abduction

I have already said that the GW-model[22] does a good job in modeling the ignorance-preserving character of abduction and—I am convinced—in designing the correct intellectual framework we should adopt especially when dealing with the problem of abduction as an inference to the best hypothesis/explanation. Following Gabbay and Wood's contention, it is clear that "[...] abduction is a procedure in which something that lacks epistemic virtue is accepted because it has virtue of another kind" (Gabbay and Woods 2005, p. 62).

For example: "Let S be the standard that you are not able to meet (e.g., that of mathematical proof). It is possible that there is a lesser epistemic standard S' (e.g., having reason to believe) that you do meet" (Woods 2013, p. 370). Focusing attention on this cognitive aspect of abduction, and adopting a logical framework centered on practical agents, Gabbay and Woods (2005) contend that abduction (basically seen as a *scant-resource* strategy, which proceeds in absence of knowledge) presents an *ignorance-preserving* (or, better, an *ignorance mitigating*) character. Of course "[...] it is not at all necessary, or frequent, that the abducer be wholly in the dark, that his ignorance be total. It needs not be the case, and typically isn't, that the abducer's choice of a hypothesis is a blind guess, or that nothing positive can be said of it beyond the role it plays in the subjunctive attainment of the abducer's original target

[22]That is Gabbay and Woods Schema.

(although sometimes this is precisely so)" (Woods 2013, p. 249). In this perspective, abductive reasoning is a *response* to an ignorance-problem: one has an ignorance-problem when one has a cognitive target that cannot be attained on the basis of what one currently knows. Ignorance problems trigger one or other of three responses. In the first case, one overcomes one's ignorance by attaining some additional knowledge (subduance). In the second instance, one yields to one's ignorance (at least for the time being) (surrender). In the third instance, one abduces (Woods 2013, Chap. 11) and so has some positive basis for new action even if in the presence of the constitutive ignorance.

From this perspective the general form of an abductive inference can be symbolically rendered as follows. Let α be a proposition with respect to which you have an ignorance problem. Putting T for the agent's epistemic target with respect to the proposition α at any given time, K for his knowledge-base at that time, K^* for an immediate accessible successor-base of K that lies within the agent's means to produce in a timely way,[23] R as the attainment relation for T, $\rightsquigarrow$ as the *subjunctive* conditional relation, H as the agent's hypothesis, $K(H)$ as the revision of K upon the addition of H, $C(H)$ denotes the conjecture of H and H^c its activation. The general structure of abduction can be illustrated as follows (GW-schema):

1. $T!\alpha$	[setting of T as an epistemic target with respect to a proposition α]
2. $\neg(R(K, T)$	[fact]
3. $\neg(R(K^*, T)$	[fact]
4. $H \notin K$	[fact]
5. $H \notin K^*$	[fact]
6. $\neg R(H, T)$	[fact]
7. $\neg R(K(H), T)$	[fact]
8. If $H \rightsquigarrow R(K(H), T)$	[fact]
9. H meets further conditions $S_1, \ldots. S_n$	[fact]
10. Therefore, $C(H)$	[sub-conclusion, 1-9]
11. Therefore, H^c	[conclusion, 1-10]

It is easy to see that the distinctive epistemic feature of abduction is captured by the schema. It is a given that H is not in the agent's knowledge-set. Nor is it in its immediate successor. Since H is not in K, then the revision of K by H is not a knowledge-successor set to K. Even so, $H \rightsquigarrow (K(H), T)$. So we have an ignorance-preservation, as required [cf. (Woods 2013, p. 370)].

[*Note*: Basically, line 9. indicates that H has no more plausible or relevant rival constituting a greater degree of subjunctive attainment. Characterizing the S_i is the

[23] K^* is an accessible successor of K to the degree that an agent has the know-how to construct it in a timely way; i.e., in ways that are of service in the attainment of targets linked to K. For example if I want to know how to spell 'accommodate', and have forgotten, then my target can't be hit on the basis of K, what I now know. But I might go to my study and consult the dictionary. This is K^*. It solves a problem originally linked to K.

most difficult problem for abductive cognition, given the fact that in general there are many possible candidate hypotheses. It involves for instance the *consistency* and *minimality* constraints.[24] These constraints correspond to the lines 4 and 5 of the standard AKM schema of abduction,[25] which is illustrated as follows:

1. E
2. $K \not\rightsquigarrow E$
3. $H \not\rightsquigarrow E$
4. $K(H)$ is consistent
5. $K(H)$ is minimal
6. $K(H) \rightsquigarrow E$
7. Therefore, H.

(Gabbay and Woods 2005, pp. 48–49)

where of course the conclusion operator $\rightsquigarrow$ cannot be classically interpreted].[26]

Finally, in the GW-schema $C(H)$ is read "It is justified (or reasonable) to conjecture that H" and H^c is its activation, as the basis for *planned* "actions".

In sum, in the GW-schema T cannot be attained on the basis of K. Neither can it be attained on the basis of any successor K^* of K that the agent knows then and there how to construct. H is not in K: H is a hypothesis that when reconciled to K produces an updated $K(H)$. H is such that if it were true, then $K(H)$ would attain T. The problem is that H is *only hypothesized*, so that the truth is not assured. Accordingly Gabbay and Woods contend that $K(H)$ *presumptively* attains T. That is, having hypothesized that H, the agent just "presumes" that his target is now attained. Given the fact that presumptive attainment is not attainment, the agent's abduction must be considered as preserving the ignorance that already gave rise to her (or its, in the case for example of a machine) initial ignorance-problem. Accordingly, abduction does not have to be considered the "solution" of an ignorance problem, but rather a response to it, in which the agent reaches presumptive attainment rather than actual attainment. $C(H)$ expresses the conclusion that it follows from the facts of the

[24] I have shown in this article that, in the case of inner processes in organic agents, this sub-process—here explicitly modeled thanks to a formal schema—is considerably implicit, and so also linked to unconscious ways of inferring, or even, in Peircean terms, to the activity of the instinct (Peirce 1931–1958, 8.223) and of what Galileo called the *lume naturale* (Peirce 1931–1958, 6.477), that is the innate fair for guessing right. This and other cognitive aspects can be better illustrated thanks to my alternative EC-model model of abduction.

[25] The classical schematic representation of abduction is expressed by what Gabbay and Woods (2005) call AKM-schema, which is contrasted to their own (GW-schema), which I am just explaining in this appendix. For *A* they refer to Aliseda (1997, 2006), for *K* to Kowalski (1979), Kuipers (1999), and Kakas et al. (1993), for *M* to Magnani (2001) and Meheus et al. (2002). A detailed illustration of the AKM schema is given in Magnani (2009, Chap. 2, Sect. 2.1.3).

[26] The target has to be an explanation and $K(H)$ bears R^{pres} [that is the relation of presumptive attainment] to T only if there is a proposition V and a consequence relation $\rightsquigarrow$ such that $K(H) \rightsquigarrow V$, where V represents a *payoff proposition* for T. In turn, in this schema explanations are interpreted in consequentialist terms. If E is an explanans and E' an explanandum the first explains the second only if (some authors further contend if and only if) the first implies the second. It is obvious to add that the AKM schema embeds a D-N (deductive-nomological) interpretation of explanation, as I have already stressed in Magnani (2001, p. 39).

schema that H is a worthy object of conjecture. It is important to note that in order to solve a problem it is not necessary that an agent actually conjectures a hypothesis, but it is necessary that she states that the hypothesis is *worthy of conjecture*.

Finally, considering H justified to conjecture is not equivalent to considering it justified to accept/activate it and eventually to send H to experimental trial. H^c denotes the *decision* to release H for further premissory work in the domain of enquiry in which the original ignorance-problem arose, that is the activation of H as a positive *cognitive* basis for action. Woods usefully observes:

> There are lots of cases in which abduction stops at line 10, that is, with the conjecture of the hypothesis in question but not its activation. When this happens, the reasoning that generates the conjecture does not constitute a positive basis for new action, that is, for acting *on* that hypothesis. Call these abductions *partial* as opposed to full. Peirce has drawn our attention to an important subclass of partial abductions. These are cases in which the conjecture of H is followed by a decision to submit it to experimental test. Now, to be sure, doing this is an action. It is an action *involving* H but it is not a case of acting *on* it. In a full abduction, H is activated by being released for inferential work in the domain of enquiry within which the ignorance-problem arose in the first place. In the Peircean cases, what counts is that H is withheld from such work. Of course, if H goes on to test favourably, it may then be released for subsequent inferential engagement (Woods 2009, p. 255).

We have to remember that this process of evaluation and so of activation of the hypothesis, is not abductive, but inductive, as Peirce contended. Woods adds: "Now it is quite true that epistemologists of a certain risk-averse bent might be drawn to the admonition that partial abduction is as good as abduction ever gets and that complete abduction, inference-activation and all, is a mistake that leaves any action prompted by it without an adequate rational grounding. This is not an unserious objection, but I have no time to give it its due here. Suffice it to say that there are real-life contexts of reasoning in which such conservatism is given short shrift, in fact is ignored altogether. One of these contexts is the criminal trial at common law" (Woods 2009, p. 255).

In the framework of the GW-schema it cannot be said that testability is intrinsic to abduction, such as it is instead maintained in the case of some passages of Peirce's writings.[27] This activity of testing, I repeat, which in turn involves degrees of risk proportioned to the strength of the conjecture, is strictly cognitive/epistemic and inductive in itself, for example an experimental test, and it is an intermediate step to release the abduced hypothesis for inferential work in the domain of enquiry within which the ignorance-problem arose in the first place.

Through abduction the basic ignorance—that does not have to be considered total "ignorance"—is neither solved nor left intact: it is an ignorance-preserving accommodation of the problem at hand, which "mitigates" the initial cognitive "irritation" (Peirce says "the irritation of doubt").[28] As I have already stressed, in a defeasible way, further action can be triggered either to find further abductions or to "solve"

---

[27] When abduction stops at line 10 (cf. the GW schema), the agent is not prepared to accept $K(H)$, because of supposed adverse consequences.

[28] "The action of thought is excited by the irritation of doubt, and ceases when belief is attained; so that the production of belief is the sole function of thought" (Peirce 1987, p. 261).

the ignorance problem, possibly leading to what the "received view" has called the *inference to the best explanation* (IBE).

It is clear that in the framework of the GW-schema the inference to the best explanation— if considered as a truth conferring achievement justified by the empirical approval—cannot be a case of abduction, because abductive inference is constitutively ignorance-preserving. In this perspective the inference to the best explanation involves the generalizing and evaluating role of *induction*. Of course it can be said that the requests of originary thinking are related to the depth of the abducer's ignorance.

In Magnani (2013) I have extensively analyzed and criticized the ignorance-preserving character of abduction, taking advantage of my *eco-cognitive model* (EC-model) of abduction and of three examples taken from the areas of both philosophy and epistemology. Indeed, through abduction, knowledge can be enhanced, even when abduction is not considered an inference to the best explanation in the classical sense of the expression, that is an inference necessarily characterized by an empirical evaluation phase, or an inductive phase, as Peirce called it. Hence, abduction is not always ignorance-preserving, but also knowledge enhancing.

Finally, let us reiterate a passage taken from Woods' quotation above: "There are lots of cases in which abduction stops at line 10, that is, with the conjecture of the hypothesis in question but not its activation. When this happens, the reasoning that generates the conjecture does not constitute a positive basis for new action, that is, for acting *on* that hypothesis". We do not have to forget that, as I have illustrated in Magnani (2013) and in the present article, various ways of *positively* enhancing knowledge are occurring also in the case of evidentially inert abductions (perception, instinct, scientific models, etc.), and very often a human abductive guess is activated and becomes a basis for action even if it has provided absolutely unreliable—if seen in the light of positive rational criteria of acceptation—knowledge. This is the case for example of the role of abductive guesses in the so-called fallacious and other kinds of reasoning, where the simple struggle that is occurring at the level of the so-called *coalition enforcement* is at stake.[29]

References

Albertazzi, L., van Tonder, G. J., & Vishwanath, D. (Eds.). (2011). *Perception beyond inference: The information content of visual processes*. Cambridge, MA: The MIT Press.

Aliseda, A. (1997). Seeking explanations: abduction in logic, philosophy of science and artificial intelligence. PhD thesis, Amsterdam: Institute for Logic, Language and Computation.

Aliseda, A. (2006). *Abductive reasoning. Logical investigations into discovery and explanation*. Berlin: Springer.

Anderson, D. R. (1987). *Creativity and the philosophy of Charles S. Peirce*. Oxford: Claredon Press.

Angius, N. (2013). Towards model-based abductive reasoning in automated software testing. *Logic Journal of the IGPL*, *21*(6), 931–942.

[29] I have analyzed the role of abduction in coalition enforcement, as a cognitive tool of the so-called *military intelligence* in Magnani (2011) and, in the case of *epistemic warfare*, in Magnani (2012).

Banerjee, B. (2006). A layered abductive inference framework for diagramming group motions. *Logic Journal of the IGPL, 14*(2), 363–378.
Brewer, W. F., & Lambert, B. L. (2001). The theory-ladenness of observation and the theory-ladenness of the rest of the scientific process. *Philosophy of Science, 68*, S176–S186 (Proceedings of the PSA 2000 Biennal Meeting).
Churchland, P. M. (1988). Perceptual plasticity and theoretical neutrality: A reply to Jerry Fodor. *Philosophy of Science, 55*, 167–187.
Cohn, A. G., Magee, D. R., Galata, G., Hogg, D. C., & Hazarika, S. M. (2002). Towards an architecture for cognitive vision using qualitative spatio-temporal representations and abduction. In C. Freska, C. Habel, & K. F. Wender (Eds.), *Spatial cognition III* (pp. 232–248). Berlin: Springer.
de Cheveigné, A. (2006). Hearing, action, and space. In D. Andler, Y. Ogawa, M. Okada, & S. Watanabe (Eds.), *Reasoning and cognition* (pp. 253–264). Tokyo: Keio University Press.
Fodor, J. (1984). Observation reconsidered. *Philosophy of Science, 51*, 23–43. Reprinted in [Goldman, 1993, pp. 119–139].
Frankfurt, H. (1958). Peirce's notion of abduction. *Journal of Philosophy, 55*, 593–397.
Gabbay, D. M., & Woods, J. (2008). *The reach of abduction*. Vol. 2 of A practical logic of cognitive systems. Amsterdam: North-Holland.
Goldman, A. I. (Ed.). (1993). *Readings in philosophy and cognitive science*. Cambridge, MA: Cambridge University Press.
Gooding, D. (1996). Creative rationality: Towards an abductive model of scientific change. *Philosophica, 58*(2), 73–102.
Hoffmann, M. H. G. (1999). Problems with Peirce's concept of abduction. *Foundations of Science, 4*(3), 271–305.
Hoffmann, M. H. G. (2004). How to get it. Diagrammatic reasoning as a tool for knowledge development and its pragmatic dimension. *Foundations of Science, 9*, 285–305.
Hutchins, E. (1995). *Cognition in the wild*. Cambridge, MA: The MIT Press.
Hutchins, E. (2005). Material anchors for conceptual blends. *Journal of Pragmatics, 37*, 1555–1577.
Kakas, A., Kowalski, R. A., & Toni, F. (1993). Abductive logic programming. *Journal of Logic and Computation, 2*(6), 719–770.
Kant, I. (1929). *Critique of pure reason*. London: MacMillan (Kemp Smith, N. Trans., originally published 1787, reprint 1998).
Kapitan, T. (1997). Peirce and the structure of abductive inference. In N. Houser, D. D. Roberts, & J. van Evra (Eds.), *Studies in the logic of Charles Sanders Peirce* (pp. 477–496). Bloomington and Indianapolis: Indiana University Press.
Kowalski, R. A. (1979). *Logic for problem solving*. New York: Elsevier.
Kuipers, T. A. F. (1999). Abduction aiming at empirical progress of even truth approximation leading to a challenge for computational modelling. *Foundations of Science, 4*, 307–323.
Mackonis, A. (2013). Inference to the best explanation, coherence and other explanatory virtues. *Synthese, 190*, 975–995.
Magnani, L. (1996). Visual abduction: Philosophical problems and perspectives. Comment to R. Lindsay, Genralizing from diagrams, In *AAAI Spring Symposium* (pp. 21–24). Stanford, CA: American Association for Artificial Intelligence.
Magnani, L. (2001). *Abduction, reason, and science. Processes of discovery and explanation*. New York: Kluwer Academic/Plenum Publishers.
Magnani, L. (2009). *Abductive cognition. The epistemological and eco-cognitive dimensions of hypothetical reasoning*. Heidelberg/Berlin: Springer.
Magnani, L. (2011). *Understanding violence. The interwining of morality, religion, and violence: A philosophical stance*. Heidelberg/Berlin: Springer.
Magnani, L. (2012). Scientific models are not fictions. Model-based science as epistemic warfare. In L. Magnani, & P. Li (Eds.), *Philosophy and cognitive science. Western and eastern studies* (pp. 1–38). Heidelberg/Berlin: Springer.

Magnani, L. (2013). Is abduction ignorance-preserving? Conventions, models, and fictions in science. *Logic Journal of the IGPL, 21*(6), 882–914.
Magnani, L., Civita, S., & Previde Massara, G. (1994). Visual cognition and cognitive modeling. In V. Cantoni (Ed.), *Human and machine vision: Analogies and divergences* (pp. 229–243). New York: Plenum Publishers.
Marcus, G. F. (2006). Cognitive architecture and descent with modification. *Cognition, 101*, 443–465.
Meheus, J., Verhoeven, L., Van Dyck, M., & Provijn, D. (2002). Ampliative adaptive logics and the foundation of logic-based approaches to abduction. In L. Magnani, N. J. Nersessian, & C. Pizzi (Eds.), *Logical and computational aspects of model-based reasoning* (pp. 39–71). Dordrecht: Kluwer Academic Publishers.
Paavola, S. (2004). Abduction through grammar, critic and methodeutic. *Transactions of the Charles S. Peirce Society, 40*(2), 245–270.
Peirce, C. S. (1931–1958). *Collected papers of Charles Sanders Peirce*. Harvard University Press, Cambridge, MA. vols. 1–6, Hartshorne, C. and Weiss, P., (Eds.); vols. 7–8, Burks, A. W., (Ed.), 1931–1958.
Peirce, C. S. (1966). *The Charles S. Peirce papers: Manuscript collection in the houghton library*. Worcester, MA: The University of Massachusetts Press. Annotated Catalogue of the Papers of Charles S. Peirce. Numbered according to Richard S. Robin. Available in the Peirce Microfilm edition. Pagination: CSP = Peirce / ISP = Institute for Studies in Pragmaticism.
Peirce, C. S. (1976). *The new elements of mathematics by Charles Sanders Peirce* (Vols. I–IV). The Hague-Paris/Atlantic Higlands, NJ: Mouton/Humanities Press (edited by C. Eisele).
Peirce, C. S. (1986). *Pragmatism as a principle and method of right thinking. The 1903 Harvard lectures on pragmatism*. Albany, NY: State University of New York Press. Ed. by Turrisi, P. A., and Peirce, C. S. *Lectures on Pragmatism*, Cambridge, MA, March 26–May 17, 1903. Reprinted in [Peirce, 1992–1998, II, pp. 133–241].
Peirce, C. S. (1987). *Historical perspectives on Peirce logic of science: A history of science* (Vols. I–II). Berlin: Mouton. (edited by C. Eisele).
Peirce, C. S. (1992–1998). *The essential Peirce. Selected philosophical writings*. Indiana University Press, Bloomington and Indianapolis, 1992–1998. Vol. 1 (1867–1893), Ed. by Houser, N. and Kloesel, C., vol. 2 (1893–1913) Ed. by the Peirce Edition Project.
Raftopoulos, A. (2001a). Is perception informationally encapsulated? The issue of theory-ladenness of perception. *Cognitive Science, 25*, 423–451.
Raftopoulos, A. (2001b). Reentrant pathways and the theory-ladenness of perception. *Philosophy of Science, 68*, S187–S189 (Proceedings of PSA 2000 Biennal Meeting).
Raftopoulos, A. (2009). *Cognition and perception. How do psychology and neural science inform philosophy*? Cambridge, MA: The MIT Press.
Roberts, L. D. (2004). The relation of children's early word acquisition to abduction. *Foundations of Science, 9*(3), 307–320.
Shanahan, M. (2005). Perception as abduction: Turning sensory data into meaningful representation. *Cognitive Science, 29*, 103–134.
Stjernfelt, F. (2007). *Diagrammatology, ontology, and semiotics. An investigation on the borderlines of phenomenology*. Berlin/New York: Springer.
Szentagothai, J., & Arbib, M. A. (1975). *Conceptual models of neural organization*. Cambridge, MA: The MIT Press.
Thagard, P. (1988). *Computational philosophy of science*. Cambridge, MA: The MIT Press.
Thagard, P. (2005). How does the brain form hypotheses? Towards a neurologically realistic computational model of explanation. In P. Thagard, P. Langley, L. Magnani, & C. Shunn, (Eds.), *Symposium "Generating explanatory hypotheses: mind, computer, brain, and world"*. Cognitive Science Society. Proceedings of the 27th International Cognitive Science Conference, CD-Rom, Stresa, Italy.

Thagard, P. (2007). Abductive inference: From philosophical analysis to neural mechanisms. In A. Feeney & E. Heit (Eds.), *Inductive reasoning: experimental developmental, and computational approaches* (pp. 226–247). Cambridge: Cambridge University Press.
Tiercelin, C. (2005). Abduction and the semiotic of perception. *Semiotica, 153*(1/4), 389–412.
Turrisi, P. A. (1990). Peirce's logic of discovery: Abduction and the universal categories. *Transactions of the Charles S. Peirce Society*, *26*, 465–497.
Woods, J. (2009). Ignorance, inference and proof: Abductive logic meets the criminal law. In G. Tuzet, & D. Canale (Eds.), *The rules of inference: Inferentialism in law and philosophy* (pp. 151–185), Milan, Egea: Bocconi University.
Woods, J. (2011). Recent developments in abductive logic. *Studies in History and Philosophy of Science*, *42*(1), 240–244 (Essay Review of L. Magnani, Abductive Cognition: The Epistemologic and Eco-Cognitive Dimensions of Hypothetical Reasoning, Heidelberg/Berlin:Springer, 2009)
Woods, J. (2013). *Errors of reasoning. Naturalizing the logic of inference*. London: College Publications.

From Visual Abduction to Abductive Vision

Woosuk Park

Abstract In order to fathom Peirce's mind, and thereby in order to do science and philosophy in Peircean way, vision seems to be a perfect point of departure. For vision allows us to rethink what true interdisciplinarity would be like in our research. In this article, I shall show the central importance of visual abduction and abductive vision in our future study of abduction as well as Peirce's thought. As exemplified well in Magnani's study of abduction, we have good reasons to go with and beyond Peirce. After briefly scheming Peirce's view on perception as abduction, I shall report what has been done in recent years in the fields of visual abduction and abductive vision. The centrality of visual abduction in Magnani's theory of manipulative abduction will be one focal point. Another will be an examination of Raftopoulos' discussion of abduction in late vision.

In order to fathom Peirce's mind, and thereby in order to do science and philosophy in Peircean way, vision seems to be a perfect point of departure. For vision allows us to rethink what true interdisciplinarity would be like in our research. In this article, I shall show the central importance of visual abduction and abductive vision in our future study of abduction as well as Peirce's thought. As exemplified well in Magnani's study of abduction, we have good reasons to go with and beyond Peirce. After briefly scheming Peirce's view on perception as abduction, I shall report what has been done in recent years in the fields of visual abduction and abductive vision. The centrality of visual abduction in Magnani's theory of manipulative abduction will be one focal point. Another will be an examination of Raftopoulos' discussion of abduction in late vision.

W. Park (✉)
KAIST, 373-1 Guseong-dong, Yuseong-gu, Daejeon, South Korea
e-mail: woosukpark@kaist.ac.kr

L. Magnani et al. (eds.), *Philosophy and Cognitive Science II*,
Studies in Applied Philosophy, Epistemology and Rational Ethics 20,
DOI 10.1007/978-3-319-18479-1_8

1 Peirce on Perception as Abduction

As is nicely addressed in Sami Paavola's influential paper, it is still controversial whether abduction is instinct or inference[1]: "If abduction relies on instinct, it is not a form of reasoning, and if it is a form of reasoning, it does not rely on instinct" (Paavola 2005, p. 131). Fortunately, Lorenzo Magnani's recent discussion of animal abduction sheds light on both instinctual and inferential character of Peircean abduction. Contrary to many commentators, who find conflicts between abduction as instinct and abduction as inference, he claims that they simply co-exist (Magnani 2009).[2] Perhaps one of the key texts for Peirce's view is the following:

> The third cotary[3] proposition is that abductive inference shades into perceptual judgment without any sharp line of demarcation between them; or in other words our first premises, the perceptual judgments, are to be regarded as an extreme case of abductive inferences, from which they differ in being absolutely beyond criticism (EP 2, p. 227).

Indeed, Magnani uses this passage as the crucial evidence for the view that "Perception is abductive in itself", thereby resolving Paavola's problem (Magnani 2009, p. 268). And, I fully agree with Magnani for this. But what exactly do we mean by this? In what context and for what purpose did Peirce and Magnani claim that perception is a special kind of abduction? What far-reaching implications are there in this apparently radical and controversial claim?.

It is interesting to note that, in his subsequent discussion immediately following this, Magnani points out that "[i]f perceptions are abductions they are basically withdrawable, just like the scientific hypotheses abductively found" [Ibid.]. Further, after discussing the semiotic and abductive character of perception with the focus on controllability, he ultimately summarizes the outcome of the discussion as follows:

> In summary, judgments in perception are fallible but indubitable abductions – we are not in any condition to psychologically conceive that they are false, as they are unconscious habits of inference (Magnani 2009, p. 269).

Similarly, Tiercelin understands what Peirce was doing in his discussion of perception as abduction as providing us with "true connecting links between abductions and perceptions, midway between a *seeing* and a *thinking*" (Tiercelin 2005, p. 393). Further, she finds Peirce as illustrating such connecting links by three different kinds of experiences: (1) optical illusions, (2) phenomena that "involve both our constitution as a natural tendency to *interpret* and some *intentional* characteristics of the objects themselves", and (3) cases where "we can repeat the sense of a conversation

---

[1] This section is based on Park (2014).

[2] Magnani (2009), especially Chap. 5 "Animal Abduction: From Mindless Organisms to Artifactual Mediators", which was originally published in Magnani and Li (2007, pp. 3–38).

[3] According to Campbell, the word "cotary" is a neologism from Latin, meaning "whetstone". So, Peirce's three cotary propositions of pragmatism are supposed to sharpen the concept of pragmatism (Campbell 2011, p. 54). I am indebted to Lorenzo Magnani for this reference. More detailed further hints are found in the editors' footnote #1 for Peirce's "Pragmatism as the Logic of Abduction". (EP, p. 530).

but we are often quite mistaken as to what words were uttered" (Tiercelin 2005, pp. 393–394; Peirce 1998, pp. 228–229).

Tiercelin's reconstruction of the historical and/or theoretical background, against which Peirce was presenting his views of perception as abduction, is even more pertinent. For, roughly speaking, she claims that Peirce's stance on perception evolved from that of emphasizing the inferential character of perception to an "immediate theory of perception". Tiercelin finds in Peirce's earliest writings (1865–1868) views reminiscent of Berkeley and Helmholtz (Tiercelin 2005, p. 390). According to her, many of Peirce's earlier views have not changed, when he presented his views on perception as abduction in 1903. "[S]ince the 1880s at least", she notes, however, Peirce adopted new ways of explaining the relations between thought and reality (Tiercelin 2005, p. 391).

The rivalry between the inferential theory of perception and the immediate theory of perception is not over yet. We can witness this fact by simply referring to recent articles such as Norman (2002), where he contrasts what he calls "the constructivist and ecological theories". In view of all this, Peirce's transition from an inferential theory to an immediate theory of perception is truly intriguing. For one might simply assume Peircean view of perception as abduction to be a kind of inferential theory of perception. Here is a good example:

> There is a long tradition of belief in philosophy and psychology that perception relies on some form of inference (Kant 1787, 1968; von Helmholtz 1967; Bruner 1957; Rock 1983; Gregory 1987; Fodor 1983). But this form of inference has been typically thought of as some form of deduction, or simple recognition, or feature-based classification, not as abduction (Josephson and Josephson 1982, p. 238).

As they point out, only in recent times and only occasionally researchers have proposed that perception involves some form of abduction. Insofar as one is preoccupied with the idea that abduction is a kind of inference, it would be extremely hard, if not impossible, to understand what is meant by "Peirce's transition from an inferential theory to an immediate theory of perception".

2 Magnani on Visual Abduction

We may understand "visual abduction" very broadly in such a way that Jon Barwise and his colleagues' research on visual Information and diagrammatic reasoning in the 1980s as a precursor to the study of visual abduction, which started in the mid 1990s. Interestingly, it seems 1994 that can be counted as the birth date for the recent study of visual abduction. For, not only Cameron Shelley's MA thesis entitled "*Visual Abductive Reasoning*" but also Lorenzo Magnani and his co-workers' published paper entitled "Visual cognition and cognitive modeling" appeared in 1994 (Shelley 1994; Magnani et al. 1994). Later, Shelley, sometimes together with his former supervisor Paul Thagard, published several important pieces on visual abduction (Shelley 1995, 1996, 2003; Thagard and Shelley 1997). Magnani also contributed several papers and books dealing with visual abduction (Magnani 2001, 2007, 2009, 2010, 2011). Recently, both Shelley and Magnani presented their most recent thoughts about visual abduction at the International Workshop on "Visual Abduction or Abductive Vision?" at Korea Advanced Institute of Science and Technology in November, 2013 (Shelley 2015; Magnani 2015). From these two presentations—see the related articles published in this volume—of Shelley and Magnani, perhaps we can extract almost everything we know about visual abduction.

In order to appreciate the importance of visual abduction for Magnani, it may be useful to take a glimpse of Magnani's multiple distinctions of abduction: (1) selective/creative; (2) theoretical/manipulative; and (3) sentential/model-based. Each of these distinctions plays a crucial role in Magnani's theory of abduction. However, as our focal interest lies in understanding the relationships between these three distinctions, let it suffice to quote one most revealing text[4]:

> What I call *theoretical abduction* certainly illustrates much of what is important in creative abductive reasoning, in humans and in computational programs, especially the objective of selecting and creating a set of hypotheses (diagnoses, causes, hypotheses) that are able to dispense good (preferred) explanations of data (observations), but fails to account for many cases of explanations occurring in science and in everyday reasoning when the exploitation of environment is crucial. ...I maintain that there are two kinds of theoretical abduction, "sentential", related to logic and to verbal/symbolic inferences, and "model-based", related to the exploitation of internalized models of diagrams, pictures, etc., cf. below in this chapter (cf. Fig. 1.2) (Magnani 2009, p. 11).

This text is important because it presents a tentative definition of theoretical abduction and the subdivision theoretical abduction into sentential and model-based abductions. It also significantly identifies theoretical abduction *as a creative abductive reasoning*. Insofar as Magnani views theoretical abduction as a kind of creative abduction, and again insofar as he tries to distinguish between theoretical and manipulative abductions, he must view manipulative abduction as a kind of creative abduction. In this regard, the following text could be more informative:

> *Manipulative abduction* (Magnani 2001) – contrasted with theoretical abduction – happens when we are thinking through doing and not only, in a pragmatic sense, about doing. ... Manipulative abduction refers to an extra-theoretical behavior that aims at creating communicable accounts of new experiences to integrate them into previously existing systems of experimental and linguistic (theoretical) practices (Magnani 2009, p. 39, 2001, p. 53).

Further, in Magnani's writings it is also stressed that manipulative abduction is occurring taking advantage of those model-based (for example, iconic) aspects that are embedded in external models:

> We have seen that manipulative abduction is a kind of abduction, usually model-based and so intrinsically "iconic", that exploits external models endowed with delegated (and often implicit) cognitive and semiotic roles and attributes (p. 58).

---

[4]Elsewhere I discussed Magnani's multiple distinctions of abduction in connection with the problem of classifying types or patterns of abduction. I am adopting my previous discussion in the following [Park (2015)].

This line of thought indicates clearly a possibility that Magnani's multiple distinctions of abduction may work in entirely different fashion, in such a way that each distinction represents a different dimension in our understanding of abduction. What I have in mind might be presented crudely as in the Cubic Model of Magnani's classification of abduction (See Fig. 1).

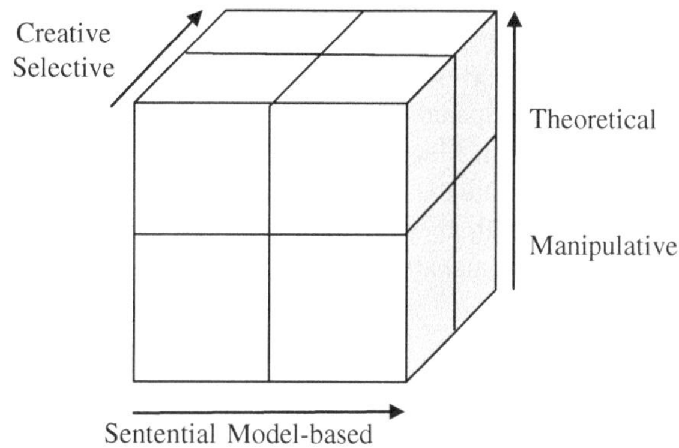

Fig. 1 Magnani's classification of abduction: Cubic Model

Now, with Magnani's multiple distinctions between different types of abduction, how are we to understand visual abduction? It seems to me that Magnani tends to approach visual abduction by invoking geometrical diagrams. Even though Magnani (2015) does not discuss geometrical diagrams extensively, Magnani already dealt with the role of model-based and manipulative abductions in geometrical reasoning in several places (Magnani 2001, 2009, 2015; Magnani and Dossena 2005). Since there have been many attempts to understand Peirce's philosophy of mathematics focusing on his distinction between corollarial and theorematic reasoning (Levy 1997; Hoffmann 2000; Sternfelt 2007, 2011), Magnani's results in visual abduction can be easily combined with previous results in diagrammatic reasoning in geometry. For example, Peicean corollarial reasoning would be mode-based, theoretical, visual abduction. On the other hand, Peicean theorematic reasoning would be model-based, manipulative, visual abduction (Magnani 2009, pp. 117–118; 176–178).

To assimilate Magnani's views of manipulative abduction to Peirce's theory of diagrammatic reasoning in geometry, the most difficult problem to overcome would be that Peircean corollarial and theorematic reasonings are deductions. Magnani claims that "theorematic deduction can be easily interpreted in terms of manipulative abduction" (Magnani 2009, p. 178). However, it is not clear what he has in mind. Probably, further hints can be secured from the following quote from Magnani:

> As I have already indicated Peirce further distinguished a "corollarial" and a "theoric" part within "theorematic reasoning", and connected theoric aspects to abduction (Hoffmann 1999, p. 293): "Thêoric reasoning [. . .] is very plainly allied to" what is normally called abduction (Peirce 1966, p. 754, ISP, p. 8; Magnani 2009, p. 181).

As Hoffmann points out, however, there can different possible interpretations of what Peirce says: "either he *identified* abduction/retroduction and theoric reasoning here or he claimed that there is abduction in mathematics beyond theoric deduction" (Hoffmann 1999, p. 293).

Be that as it may, it must be one of the most significant achievements in Magnani's study of abduction that he quite successfully identified and highlighted a category of manipulative abduction. As times go by, he tends to emphasize its significance more and more. It seems that he even counts manipulativeness as an inherent feature of all abduction. Probably, here Magnani is going beyond Peirce. For, as far as I know, Peirce never suggested explicitly such a category as "manipulative abduction". Even if Peirce extensively discussed geometrical diagrammatic reasoning, and thereby implicitly anticipated whatever Magnani has to say about its manipulative character, we should credit Magnani for the discovery of manipulative abduction. Now, what I want to emphasize is that Magnani's discovery of manipulative abduction may have started with visual abduction in diagrammatic reasoning. If this observation is correct, then visual abduction is the core of Magnani's theory of abduction.

3 Peirce as a Psychologist

Though rarely highlighted, there seems to be no doubt that Peirce was one of the leading American experimental psychologists in the late nineteenth and the early twentieth century.[5] Let us find the birth date of experimental psychology in "1879", when Wilhelm Wundt established his lab at Leipzig. And let us ask what Peirce as a psychologist was doing before and after the birth of experimental psychology. Relying on Thomas C. Cadwallader's study, we may answer this question rather straightforwardly. According to Cadwallader, there were three different approaches to psychology in the mid nineteenth century: (1) the philosophical approach in Europe, which follows the Cartesian tradition of relating physiology to psychology, (2) phrenology in America, which "was largely popular and had little direct input into academic psychology", and (3) the dominant psychology in America, which was "a blend of theology and philosophy" (Cadwallader 1975, p. 168). Also, following the lead of Cadwallader, we can trace Peirce's interest in psychology back to his teen age period, for there are many interesting items in his early notebooks that deal with psychology (Cadwallader 1975, pp. 168–170). Cadwallader also observes that after graduating Harvard in 1859 Peirce gave in his writings "an increasing focus on psychological topics" (Cadwallader 1975, p. 169). What is important is that in this period Peirce began to criticize introspective psychology as untrustworthy in some of his famous work.[6] More importantly, Peirce criticized severely the British tradition of faculty

[5] This section is again drawn from Park (2014).

[6] Cadwallader cites "On a New List of Categories" (1867) [CP 1.545–1.559] and "Questions Concerning Certain Faculties Claimed for Man" (1868) [CP 5.213–5.263] in this regard (Cadwallader 1975, pp. 170–171).

psychology in his 1869 review of Noah Porter's book *The Human Intellect* (1868) "for failing to follow the lead of Wundt" (Cadwallader 1975, p. 171).[7]

It is tempting to dwell further on Peirce's particular achievements as an empirical psychologist. For example, we may want to uncover the Peircean heritage in psychology at Johns Hopkins even before G. Stanley Hall founded the psychology laboratory there (Cadwallader 1975, p. 176; Leary 2009; Green 2007). Indeed, Cadwallader enumerates several prominent psychologists "as students and/or members of his Metaphysical Club": Jastrow, John Dewey, J. McKeen Cattell, and Christine Ladd Franklin. Or, we may want to have an overview of what Peirce wrote about topics in psychology, whether it be the Bezold-Brücke phenomenon, which Cadwallader suggests to call Bezold-Peirce-Brücke phenomenon (Cadwallader 1975, p. 172), or the problem of habit, which Peirce calls "the very market place of psychology". (7.367; Cadwallader 1975, p. 175).

4 Abductive Vision: Raftopoulos

Now, I suggest to understand "visual abduction" as meaning any kind of abductive reasoning related to vision. Both Shelley's study of visual analogies in archaeology and Magnani's study of manipulative abduction in geometrical proofs fit perfectly with this understanding of "visual abduction". On the other hand, "abductive vision" is different from "visual abduction" in that it aims at uncovering the mechanism of vision itself in terms of abduction. To the best of my knowledge, the expression "abductive vision" was not introduced until the KAIST International Workshop. Also, it was at this Workshop, when we come to have the first serious attempt to reveal the mechanism of vision in terms of abductive inference. For, there Athanassios Raftopoulos read a paper entitled "Abductive Inference in Late Vision" (See also Raftopoulos 2009, 2015).

As I noted elsewhere (Park 2014), Magnani's own position regarding perception does fit quite well with our characterization of Peirce's transition from inferential theory to an immediate theory of perception. After having confirmed the strong tie between perception and reification, he appeals to Raftopoulos's recent assessment of Fodor-Churchland controversy:

> in humans perception (at least in the visual case) is not strictly modular, like Fodor (1984) argued, that is, it is not encapsulated, hardwired, domain-specific, and mandatory. Neither is it wholly abductively "penetrable" by higher cognitive states (like desires, beliefs, expectations, etc.), by means of top-down pathways in the brain and by changes in its wiring through perceptual learning, as stressed by Churchland (1988) (Magnani 2009, p. 301; Raftopoulos 2001).

[7] Based on Peirce's own recollection and the evidence from the large set of notes that began aroud 1865 (Ms. 1956), Cadwallader notes that "[a]s the 60s progressed, Wundt's influence began to be apparent in Peirce's writings". Also, based on a large notebook (Ms. 1156), Cadwallader reports that Peirce showed continued interest in Wundt by referring to Wundt's Physiological Psychology of 1874 at least 47 times (Cadwallader 1975, p. 171).

Magnani believes, even if we allow "a substantial amount of information which is theory-neutral" and "a certain degree of theory-ladenness" in perceptual learning, "this fact does not jeopardize the assumption concerning the basic cognitive impenetrability of perception":

> in sum, perception is informationally "semi-encapsulated", and also semi-hardwired, but, despite its top-down [sic] character, it is not insulated from "knowledge" (Magnani 2009, p. 301).

It is simply beyond my ability to fully grasp what Raftopoulos has achieved on the borderline between philosophy and psychology. But, as least, I can see that Raftopoulos' move locating abductive inferences in late rather than early vision is extremely interesting. Charles S. Peirce is undoubtedly the first one who addressed the problem of abductive vision. It seems highly likely that he would place abduction not only in late vision but also in early vision. Was Peirce simply mistaken because he was lacking 21st century's advanced knowledge of vision science and neuroscience? Or, after all these years, do we still have to learn something from Peirce on abductive vision?

Let us check first on what ground Raftopoulos locates abductive inference in late vision. In referring to the problem as to what the logical status of the state transformations or transitions that occur in late vision is, he writes:

> I argue that in late vision an abduction or 'inference' to the best explanation allows the construction of a representation that best fits a scene by forming hypotheses concerning the identity of object in the visual scene and eliminating rival candidates until the best fit is found (Raftopoulos, Ibid.).

Interestingly, Raftopoulos is anxious to deny that this abductive inference in late vision is discursive:

> I also defend the thesis that the abductive inferences of late vision are not instances of discursive abductive inference because they do not effect transitions from propositionally structured premises to recognitional beliefs (Ibid.).

In other words, he is quite convinced that "late vision has an irreducible visual ingredient that makes it different from discursive understanding".

Now, Raftopoulos never stated that early vision is not an abductive process. What he argues for is merely the thesis that late vision is an abductive process that is not a discursive inference. In other words, Raftopoulos has not yet discussed early vision and its relation to abduction.[8] I believe that we can extend the discussion to early vision. Further, I believe that the possibility of early abductive vision would become more arguable, if we criticize some of Raftopoulos's assumptions in his subtle and painstaking reasoning. First, he simply equates abduction with inference to the best explanation. Secondly, he simply views abduction as an inference without considering the viability of Peircean view of perception as abduction. The latter

[8] Here I am indebted to an anonymous reviewer, who convinced me that Raftotpoulos never denies the possibility of early abductive vision.

assumption was already challenged in Sects. 1 and 3. Given Peirce's intriguing treatment of perception as abduction, and the qualification of Peirce as an experimental psychologist, it seems unfair to ignore the possibility of abduction in early vision. It is highly likely that it the matter of empirical study by psychologist whether there is abduction in early vision. Even then, we should grant a fair hearing to Peirce's views taking seriously the possibility of abduction in early vision.

5 Abduction Is Not IBE

Now let us turn to Raftopoulos' first assumption.[9] In the huge literature on IBE, Raftopoulos is not the only one, who equates abduction with IBE. Starting with Gilbert Harman's influential article "Inference to the Best Explanation" (Harman 1965), it culminates at Peter Lipton's book devoted to IBE (Lipton 1991, 2004). Harman didn't bother with the possible differences between abduction and IBE. For, he states as if there are merely terminological differences between them:

> "The inference to the best explanation" corresponds approximately to what others have called "abduction," "the method of hypothesis," "hypothetic inference," "the method of elimination," "eliminative induction," and "theoretical inference" (Harman 1965, pp. 88–89).[10]

Lipton is not different from Harman in this regard (Lipton 1991, p. 58; Lipton 2004, p. 57). The textbooks or reference works of philosophy of science, with very good reasons, simply follow suit of such eminent debates among leading philosophers in equating abduction with IBE (Ladyman 2002, p. 47).

However, there have been some attempts to revolt against such a trend. Campos criticizes Lipton's appropriation as "inaccurate" on the ground that "it is not based on any systematic comparison of these concepts" (Campos 2011, pp. 419–420). Further, Gerhard Minnameier clearly contrasts abduction with IBE in terms of their functions. While abduction is for the generation of theories, IBE is for their evaluation:

> Peirce characterizes abduction as the only type of inference that is *creative* in the sense that it leads to new knowledge, especially to (possible) theoretical explanations of surprising facts. As opposed to this, IBE is about the acceptance (or rejection) of already established explanatory suggestions. Thus, while abduction marks the process of generating theories – or, more generally, concepts – IBE concerns their evaluation. However, if this is so, then

[9]This section is a kind of summary of Sect. 4 of Park (2015), where much fuller exposition was presented.

[10] Cf. Atocha Aliseda's interesting comments: "On the other hand, some authors take induction as an instance of abduction. Abduction as *inference to the best explanation* is considered by Harman [Har65] as the basic form of non-deductive inference, which includes (enumerative) induction as a special case. This confusion returns in artificial intelligence. 'Induction' is used for the process of learning from examples—but also for creating a theory to explain the observed facts [Sha91]. Thus making abduction an instance of induction. Abduction is usually restricted to producing abductive explanations in the form of facts. When the explanations are rules, it is regarded as part of induction". (Aliseda 2006, p. 34).

> both inferential types relate to entirely different steps in the process of knowledge acquisition (and, as I also try to show, of Knowledge application) (Minnameier 2004, pp. 75–76).

According to Minnameier's rather persuasive explanation, the later Peirce realized that previously he "more or less mixed up Hypothesis and Induction" (CP 8.221, 1910) (Minnameier 2004, p. 78). There are at least two interesting points that enforced Peirce to change his mind. On the one hand, he came to realize that what he had called previously "hypothesis" is rather a variant of induction. So, he even renamed "hypothesis" as "qualitative induction" [cf. NEM III/2, pp. 874, 1909; Minnameier (2004), p. 78]. On the other hand, as Minnameier pins down, "what used to be called "induction" in the sense of leading from facts to a theory about those facts would now have to be regarded as an "abductive" inference" [Ibid.]. He seems to find the ground for such an interpretation from the late Peirce's realization that induction "never can originate any idea whatever. Nor can deduction. All the ideas of science come to it by the way of Abduction" [CP 5.145, 1903; Minnameier (2004), p. 78]. Many philosophers seem to agree with Minnameier in distinguishing abduction from IBE by their primary functions: i.e., abduction as generating theories from IBE as evaluating them (Magnani 2009, p. 18; Campos 2011, p. 420; Magnani 2015, pp. 976–977).

If so, we can safely conclude that, at least in one important sense, abduction and IBE are clearly distinguished: abduction is for generation of hypotheses or theories, while IBE is for evaluating them. This conclusion might have far-reaching implications. For example, it could cause a serious trouble for Schurz and his project of classifying patterns of abduction. Since he identified without argument abduction with IBE, it could be the case that what he classifies is not the patterns of abduction but the patterns of IBE.

Another important argument against equating abduction with IBE is that IBE cannot be an abduction in GW-Model. In this regard, Magnani's keen observations on the GW- model of abduction is quite helpful:

> It is clear that in the framework of the GW-schema the inference to the best explanation– if considered as a truth conferring achievement justified by the empirical approval – cannot be a case of abduction, because abductive inference is constitutively ignorance-preserving. In this perspective the inference to the best explanation involves the generalizing and evaluating role of *induction* (Magnani 2015, pp. 5–6).

I think that I supplied enough ground to challenge Raftopoulos' assumption that abduction is IBE. If I am successful in casting doubts as to Raftopoulos' two fundamental assumptions, we seem to have ample ground to extend the possibility of abduction to early vision.

6 Concluding Remarks

The key for understanding both visual abduction and abductive vision can be found in the double aspect of abduction, i.e., abduction as instinct and abduction as inference. Given this key, we can expect to get interesting results on visual abduction by continuing and expanding what Magnani, Thagard, Shelley, and others have done. As was clear from the case of diagrammatical reasoning in geometry, we will continue to be inspired by Peirce in this venture.

As we saw above, Raftopoulos does not discuss the possibility of early abductive vision. We can surmise, however, Peirce could have located abduction even at the early vision. Peirce's scattered remarks on optical illusion can be the treasure house for us to examine this hypothesis. If so, Peirce is not merely the first but also one of the best among us who scrutinize the mysteries of visual abduction and abductive vision.

Acknowledgments This paper was first presented at KAIST International Workshop "Visual Abduction or Abductive Vision?" (November, 2013, Daejeon, Korea). A revised version was also read at Peirce Centennial Conference (July, 2014, Lowell, U.S.A.). I am enormously indebted to the suggestions and criticisms of the participants of both, especially to Lorenzo Magnani, Cameron Shelley, and Athanassios Raftopoulos. I am also grateful to the anonymous reviewers and John Woods, who saved me from some fatal mistakes.

References

Aliseda, A. (2006). *Abductive reasoning*. Dordrecht: Springer.

Bruner, J. S. (1957). On perceptual readiness. *Psychological Review, 64*(2), 123–152.

Cadwallader, T. C. (1975). Peirce as an experimental psychologist. *Transactions of the Charles S. Peirce Society, 11*, 167–186.

Campos, D. G. (2011). On the distinction between Peirce's abduction and Lipton's inference to the best explanation. *Synthese, 180*, 419–442.

Campbell, P. L. (2011), *Peirce, Pragmatism, and the right way of thinking*. Albuquerque, New mexico: Sandia National Laboratories.

Churchland, P. M. (1988). Perceptual plasticity and theoretical neutrality: A reply to Jerry Fodor. *Philosophy of Science, 55*, 167–187.

Fodor, J. (1983). *The modularity of mind*. Cambridge, MA: MIT Press.

Fodor, J. (1984). Observation reconsidered. *Philosophy of Science, 51*, 23–43.

Green, C. D. (2007). Johns Hopkins's first professorship in philosophy: A critical pivot point in the history of American psychology. *American Journal of Psychology, 120*(2), 303–323.

Gregory, R. L. (1987). Perception as hypotheses. In R. L. Gregory (Ed.), *The oxford companion to the mind* (pp. 608–611). New York: Oxford University Press.

Harman, G. (1965) The inference to the best explanation. *Philosophical Review, 74*, 88–95.

Hoffmann, M. H. G. (1999). Problems with Peirce's concept of abduction. *Foundations of Science, 4*(3), 271–305.

Josephson, J., & Josephson, S. (Eds.). (1982). *Abductive inference*. New York: Cambridge University Press.

Kant, I. (1787, 1968). *Critiques of pure reason*, (Norman Kemp Smith, Trans.). New York: St. Martins Press.

Ladyman, J. (2002). *Understanding philosophy of science*. London/New York: Routledge.

Leary, D. E. (2009). Between Peirce (1878) and James (1898): G. Stanley Hall, the origins of pragmatism, and the history of psychology. *Journal of the History of the Behavioral Sciences, 45*(1), 5–20.

Levy, S. H. (1997). Peirce's theorematic/corollarial distinction and the interconnections between mathematics and logic. In N. Houser, D. D. Roberts & J. Evra (Eds.), *Studies in the logic of Charles Sanders Peirce*, Bloomington and Indianapolis: Indiana University Press.

Macknis, A. (2013). Inference to the best explanation, coherence and other explanatory virtues. *Synthese*, *190*, 975–995.
Magnani, L. (2001). *Abduction, reason, and science: Processes of discovery and explanation*. New York: Kluwer.
Magnani, L. (2007). Animal abduction. from mindless organisms to artifactual mediators. In L. Magnani & P. Li (Eds.), *Model-based reasoning in science, technology, and medicine, studies in computational intelligence* (Vol. 64, pp. 3–37). Berlin/New York: Springer.
Magnani, L. (2009). *Abductive cognition, The epistemological and eco-cognitive dimensions of hypothetical reasoning*. Berlin: Springer.
Magnani, L. (2010). Mindless abduction: From animal guesses to artifactual mediators. In M. Bergman, S. Paavola, A.-V. Pietarinen & H. Rydenfelt (Eds.), *Ideas in Action: Proceedings of the applying Peirce Conference* (pp. 224–238). Nordic Pragmatism Network: Helsinki.
Magnani, L. (2011). Is instinct rational? Are animals intelligent?: An abductive account. In L. Carlson, C. Hoelscher & T. F. Shipley (Eds.), *Proceedings of the33rdAnnual Conference of the Cognitive Science Society* (pp. 150–155). Austin, TX: Cognitive Science Society.
Magnani, L. (2015). Understanding visual abduction. The need of the eco-cognitive model. This volume, pp. 118–139.
Magnani, L., & Dossena, R. (2005). Perceiving the infinite and the infinitesimal world: Unveiling and optical diagrams in mathematics. *Foundations of Science*, *10*, 7–23.
Magnani, L., & Li, P. (Eds.), (2007). *Model-based reasoning in science, technology, and medicine*. Springer: Berlin.
Magnani, L., Civita, S., & Massara, G. P. (1994). Visual cognition and cognitive modeling. In V. Cantoni (Ed.), *Human and machine vision: Analogies and divergencies* (pp. 229–243). New York: Plenum Press.
Minnameier, G. (2004). Peirce-suit of truth—Why inference to the best explanation and abduction ought not to be confused. *Erkenntnis*, *60*, 75–105.
Lipton, P. (1991). *Inference to the best explanation*. London/New York: Routledge.
Lipton, P. (2004). *Inference to the best explanation* (2nd ed.). London/New York: Routledge.
Norman, J. (2002). Two visual systems and two theories of perception: An attempt to reconcile the constructivist and ecological approaches. *Behavioral and Brain Sciences*, *25*, 73–144.
Paavola, S. (2005). Peircean abduction: Instinct or inference? *Semiotica*, *153*(1/4), 131–154.
Park, W. (2014). *How to learn abduction from animals?: From Avicenna to Magnani*. In L. Magnani (Ed.), *Model-based reasoning in science and technology: Theoretical and cognitive issues*, Heidelberg/Berlin: Springer.
Park, W. (2015). On classifying abduction. *Journal of Applied Logic*, *13*(3), 215–238.
Peirce, C. S. (1966). *The Charles S. Peirce papers: manuscript collection in the Houghton Library*. Worcester, MA: The University of Massachusetts Press.
Peirce, C. S. (1976). *The new elements of mathematics* (Vol. 4). In C. Eisele (Ed.), Berlin/New York: Mouton de Gruyter; Atlantic Highlands, NJ: Humanities Press [Abbreviated as NEP].
Peirce, C. S. (1998). *The essential Peirce: Selected philosophical writings* (Vol. 2). In N. Houser, & C. Kloesel (Eds.), Bloomington and Indianapolis: Indiana University Press [Abbreviated as EP].
Raftopoulos, A. (2001). Is perception informationally encapsulated? The issue of the theorylandenness of perception. *Cognitive Science*, *25*, 423–251.
Raftopoulos, A. (2009). *Cognition and perception: How do psychologt and neural science inform philosophy?*. Cambridge, Mass: The MIT Press.
Raftopoulos, A. (2015). Abductive inference in late vision. This volume, pp. 155–176.
Rock, I. (1983). *The logic of perception*. Cambridge, Mass: The MIT Press.
Shelley, C. (1994). Visual abductive reasoning, Thesis. Waterloo, On, Canada, M.A: University of Waterloo.
Shelley, C. (1995). *Visual abduction in anthropology and archaeology* (pp. 155–159). AAAI Technical Report SS-95-03.
Shelley, C. (1996). Visual abductive reasoning in archaeology. *Philosophy of Science*, *63*, 278–301.

Shelley, C. (2003). *Multiple analogies in science and philosophy*. Amsterdam/Philadelphia: John Benjamins Publishing Company.
Shelley, C. (2015). Biomorphism and models in design. This volume, pp. 209–221.
Stjernfelt, F. (2007). *Diagrammatology. An investigation on the borderlines of phenomenology, ontology, and semiotics*. Berlin: Springer.
Stjernfelt, F. (2011) Peirce's notion of diagram experiment: corollarial and theorematical experiments with diagrams. In R. Heinrich, E. Nemeth, W. Pichler & D. Wagner (Eds.), *Image and imaging in philosophy, science and the arts* (Vol. 2, pp. 305–340). Frankfurt: Ontons Verlag.
Thagard, P., & Shelley, C. (1997). Abductive reasoning: logic, visual thinking, and coherence. In M. L. Dalla Chiara, et al. (Eds.), *Logic and scientific methods* (pp. 413–427).
Tiercelin, C. (2005). Abduction and the semiotics of perception. *Semiotica, 153*, 389–412.
von Helmholtz, H. (1967). *Handbuch der physiologischen Optik*. Leipzig: Leopold Voss.

Abductive Inference in Late Vision

Athanassios Raftopoulos

Abstract In earlier work (Raftopoulos 2009), I analyzed early vision, which I claimed is a cognitively impenetrable (CI) stage of visual processing. In contradistinction, late vision is cognitively penetrated (CP) and involves the modulation of processing by cognitively driven attention. Its stages have hybrid contents, partly conceptual contents, and partly iconic analogue contents. In this chapter, I examine the processes of late vision and discuss whether late vision should be construed as a perceptual stage or as a thought-like stage. Using Jackendoff 's (1989) distinction between visual awareness and visual understanding, I argue that the contents of late vision belong to visual awareness. In late vision an abduction or "inference" to the best explanation allows the construction of a representation that best fits a scene. Given the sparse retinal image that underdetermines both the distal object and the percept, the visual system fills in the missing information to arrive at the best explanation, that is, the percept that best fits the retinal information. I argue that late vision does not consist in propositional structures formed in cognitive areas and participate in discursive reasoning and inferences, and does not implicate discursive abductive inferences from propositionally structured premises to recognitional beliefs.

1 Introduction

In earlier work (Raftopoulos 2009, 2013), I analyzed early vision, which, I have claimed, is a pre-attentional visual stage unaffected by top-down conceptual modulation; early vision is a cognitively impenetrable stage of visual processing that has nonconceptual, non-propositional content cognitively penetrated in that its processes are modulated by cognitively driven, endogenous attention. Its states have partly conceptual contents and partly visual, iconic contents, where concepts are constant, context independent, that figure constitutively in propositional contents; they correspond to lexical items.

A. Raftopoulos (✉)
Department of Psychology, University of Cyprus, Nicosia, Cyprus
e-mail: raftop@ucy.ac.cy

L. Magnani et al. (eds.), *Philosophy and Cognitive Science II*,
Studies in Applied Philosophy, Epistemology and Rational Ethics 20,
DOI 10.1007/978-3-319-18479-1_9

Here, I examine the processes of late vision and discuss (a) whether late vision should be construed as a perceptual stage or as a thought-like stage, and (b) what is the logical status of the state transformations or transitions that occur in late vision. With respect to the first problem, using Jackendoff's (1989) distinction between visual awareness and visual understanding, I argue that the states of late vision, their partly conceptual nature notwithstanding, are perceptual states properly speaking and are characterized by visual awareness. Concerning the second problem, I argue that in late vision an abduction or 'inference' to the best explanation allows the construction of a representation that best fits a scene by forming hypotheses concerning the identity of object in the visual scene and eliminating rival candidates until the best fit is found. Given the sparse retinal image that underdetermines both the distal object and the percept, the visual system fills in the missing information to arrive at the best explanation, that is, the percept that best fits the retinal information.

Philosophers and Psychologists have argued that this function of the visual system means that the visual system performs an inference that is alike the inferences performed by cognition in the space of reasons and, in this sense, perception is more like thought than a sensory process. Spelke (1988, p. 458), for example, claims that "perceiving objects may be more akin to thinking about the physical world than to sensing the immediate environment," a view that she shares with Bruner and Goldman (1947) "perception is a process of categorization in which organisms move inferentially from cues to category identity". These inferences, which I call discursive, are transitions from propositionally structured states, the premises of the inference, to a propositionally structured state, the conclusion of the inference. This transition is effected through a set of inferential rules. These rules may be either explicitly or implicitly represented in the system and looked up by viewers when they apply the rules, in which case the regularities they express are represented in the system, or they may be hardwired in the system, in which case their contents can be viewed either as constituting tacit representational knowledge, or the rules may be construed as merely performing causal state transformations without having any representational contents.

I argue that in late vision the percept is constructed by means of a set of abductive inferences from sensory information and object knowledge stored in memory to the object that best fits this information. I also defend the thesis that the abductive inferences of late vision are not instances of discursive abductive inference because they do not effect transitions from propositionally structured premises to recognitional beliefs. I also argue that there are no rules of inference stored in the system that are looked up by the system in order for the transition between states to take place. Instead, the transitions are effected through a set of hardwired computational processors that transform states to other states so that the transformations be sensitive to the regularities of the physical environment and its geometry.

In Sect. 2, I discuss late vision and its processes and explain the nature of the hybrid contents formed in it. I argue that the processes of late vision realize abductions that are not discursive inferences, and that there are not inference rules stored anywhere in the system and looked-up when perceptual state transformation are made. The visual system probably relies on pattern matching processes rather than discursive inferences to construct the percept. In the concluding section, I present connectionist models that simulate the perception of ambiguous figures as an example of how the percept may be formed through pattern matching.

2 Late Vision

The conceptually modulated stage of visual processing is called late vision. Starting at 150–200 ms, signals from higher executive centers including mnemonic circuits intervene and modulate perceptual processing in the visual cortex and this signals the onset of global recurrent processing (GRP). Specifically, in 50 ms low spatial frequency (LSF) information reaches the IT and in 100 ms high spatial frequency (HSF) information reaches the same area (Kihara and Takeda 2010). Within 130 ms post-stimulus, parietal areas in the dorsal system but also areas in the ventral pathway (IT cortex) semantically process the LSF information and determine the gist of the scene based on stored knowledge that generates predictions about the most likely interpretation of the input, even in the absence of attention. This information reenters the extrastriate visual areas and modulates (at about 150 ms) perceptual processing facilitating the analysis of HSF, by specifying certain cues in the image that might facilitate target identification (Barr 2009; Kihara and Takeda 2010). Determining the gist may speed up the fast forward sweep (FFS) of HSF by allowing faster processing of the pertinent cues, using top-down connections to preset neurons coding these cues at various levels of the visual pathway (Delmore et al. 2004). At about 150 ms hypotheses about the identity of the object(s) in the scene are formed using HSF information and information from visual working memory (WM). The hypotheses are tested against the detailed iconic information stored in early visual circuits including V1. Indeed, ERP's waveforms that distinguish scenes and objects in object recognition tasks are registered at about 150 ms in extrastriate areas and are thought to be early indices of P3 (Fabre-Thorpe et al. 2001; Johnson and Olshausen 2005).[1] Successful testing leads to the recognition of the object(s) in the visual scene. This occurs, as signaled by P3 at about 300 ms in the IT cortex, whose neurons contribute to the integration of LSF and HSF information.

A detailed analysis of the form that the hypothesis testing might take is provided by Kosslyn (1994). Note that one need not subscribe to some of the assumptions presupposed by Kosslyn's account, but these disagreements do not undermine the framework. Suppose that one sees an object. A retinotopic image is formed in the visual buffer, which is a set of visual areas in the occipital lobe that are organized retinotopically. An attentional window selects the input from a contiguous set of points for detailed processing. This is allowed by the spatial organization of the visual

[1] The P3 waveform is elicited at 300–600 ms and is generated in frontal/central, central/parietal, parietal/occipital areas, the temporal lobe, the temporal/parietal junction, and neighboring neocortical regions. The generating sites and timing onset show that P3 is associated with semantic processing and with the subjects' reports. P3 is thought to signify the consolidation of a representation in working memory.

buffer. The information included in the attention window is sent to the dorsal and ventral system where different features of the image are processed. The ventral system retrieves the features of the object, whereas the dorsal system retrieves information about the location, orientation, and size of the object. Eventually, the shape, the color, and the texture of the object are registered in anterior portions of the ventral pathway. This information is transmitted to the pattern activation subsystems in the IT cortex where the image is matched against representations stored there, and the compressed image representation of the object is thereby activated. This representation provides feedback to the visual buffer where it is matched against the input image to test the hypothesis against the fine pictorial details registered in the retinotopical areas of the visual buffer.

If the match is satisfactory, the category pattern activation subsystem sends the relevant pattern code to WM, where the object is tentatively identified (this is an hypothesis regarding the identity of an object) with the help of information arriving at the WM through the dorsal system (information about, size, location, and orientation). This hypothesis is tested against other information in WM, including semantic information about the putative object. Occasionally the match in the pattern activation subsystems is enough to select the appropriate representation in WM. On other occasions, the input to the ventral system does not match well a visual memory in the pattern activation subsystems, or the initial hypothesis formed in WM does not fit in with the other information activated in WM pertaining to the putative object (for example, the putative object does not fit in the gist of the scene). Then, another hypothesis is formed in WM. This hypothesis is tested with the help of other subsystems (including cognitive ones) that access representations of such objects and highlight their more distinctive feature. The information gathered shifts attention to a location in the image where an informative characteristic can be found. The attention window zooms on object's distinctive feature, and the pattern code for it is sent to the pattern activation subsystem and to the visual buffer where a second cycle of matching commences. ERP experiments registering the time onset of various waveforms related to specific processes in the brain confirm this analysis. The N2 component that signifies cognitively driven spatial–attentional effects on the extrastriate cortex is registered at about 170–200 ms; thus, by 170 ms spatial attention modulates visual processing.

As we saw, the object recognition system forms hypotheses regarding the identity of objects in a visual scene. For the subject's confidence, however, to reach the threshold that will allow them to form beliefs about the identity of the objects and report them, these hypotheses must be tested (Treisman 2006). When, after some hypothesis testing, the object O in the visual scene is recognized as a an F through the synergy of visual circuits and WM, the explicit belief "O is F" is formed. This occurs after 300 ms, when the viewer consolidates the object in WM and identifies it with enough confidence to report it.

The processes of late vision rely on recurrent interactions with areas outside the visual stream. This set of interactions is called *Global Recurrent Processing* (GRP). In GRP, standing knowledge, i.e., information stored in the synaptic weights is activated and modulates visual processing that up to that point was conceptually encapsulated.

During GRP the conceptualization of perception starts and the states formed have partly conceptual and eventually propositional contents. Thus, late vision involves a synergy of perceptual bottom-up processing and top-down processing, where knowledge from past experiences guides the formation of hypotheses about the identity of objects. This is the stage where the 3D sketch (that is, the representation of an object independently of the viewer's perspective) is formed. This recovery cannot be purely data-driven since what is regarded as an object depends on the subsequent usage of the information and thus depends on the knowledge about objects. Seeing 3D sketches is an instance of amodal completion, i.e., the representation of object parts that are not visible from the viewer's standpoint. In amodal completion, one does not have a perceptual impression of the object's hidden features since the perceptual system does not fill in the missing features as it happens in modal perception; the hidden features are not perceptually occurrent.

This is the proper place to examine the nature of the contents of the states of late vision, which are partly conceptual contents, partly visual contents. Late vision is the stage of visual processing in which GRP occurs, which means that the relevant processes involve not only the visual brain but also the cognitive centers of the brain (parietal cortex, prefrontal cortex, frontal cortex, etc.). Therefore, a typical state in late vision consists of neurons distributed across wide areas of the brain, whose activity is synchronized owing to the GRP; such a typical state involves neurons both in visual areas and in the cognitive centers. These neurons depending on their location encode different sorts of information.

Suppose, for example, that one of the hypotheses tested in order for viewers to recognize an object in their perceptual field is that this object is a tiger. The neurons in cognitive areas that encode the representation of tiger and some of its characteristics properties, including semantic properties, that is the neurons that store the object knowledge associated with tigers are activated. This hypothesis is tested against the iconic information stored in the sensory iconic memory mainly in the early visual areas. The testing occurs through the enhancement of the activation of the neurons in the visual areas that store incoming information from the location in the visual scene at which most likely a tiger may be found, an enhancement that is due to the top-down modulation of the visual areas from signals emanating from the cognitive states that encode information about tigers. Thus, the state of a viewer who tests this hypothesis comprises both conceptual representations that are propositionally structured and are formed in cognitive areas, and visual representations that are formed in visual areas and concern the iconic information that is relevant to the hypothesis testing. When the object is identified and a perceptual belief is formed, the object is seeing as a tiger.

Tye (2009, p. 210) claims that "states of seeing-as and seeing-that do not count as visual experiences. They are, rather, hybrid states involving visual experiences plus appropriate conceptual attitudes." One could add that the hybrid states of late vision involve both visual elements and conceptualizations of the perceptual attributives. The visual elements consist in the contextual perceptual attributives (the perceived properties) used for determining reference to objects, and in the singular elements in perception (the objects of perception). The perceptual attributives are responsible for the attributions of perception, that is, both for determining the object of perception and the assignment of properties to it. They are contextual in that they are always tied to the object of perception and, thus, they always appear and function within a particular visual context. In other words, the attributive elements/properties guide the contextual reference to the singular elements/objects since the referent in a demonstrative perceptual reference is fixed through the properties of the referent as the latter are represented in perception.

The conceptual content of the states of late vision consists in the conceptualizations of the perceptual attributives and of the singular elements in perception. When late vision begins, visual information has already spread to all brain areas and, thus, this information is available to the cognitive centers of the brain. Block (2007, pp. 320–321) calls this access cognitive access consciousness (CAC) and thinks that CAC is phenomenal consciousness plus reflection. Specifically, CAC is phenomenality plus reflection on the phenomenality that is, "phenomenality plus another state, one that is about the phenomenal state." (Block 2007 , p. 320). CAC presupposes that subjects have a state that is about their own experience and, thus, that the cognitive centers have access to the content of the experience. When the representational content of a perceptual state is accessed for cognitive reasons, this content becomes the content of a thought that the viewer can entertain simply by being in that state. Thus, cognitive access conscious states have contents that are contents of perceptual thoughts or beliefs; the contents of cognitive access conscious states are necessarily conceptual contents because only these can be the contents of thoughts. Furthermore, the concepts that enter the contents of beliefs must be concepts that the persons who entertain the belief grasp, otherwise they could not entertain the thoughts.

It should be stressed that a viewer need not be aware that they are in such a state. In other words, the content of the state, the thought, may be an implicit thought. It follows that CAC content may be content of which a subject is unaware in the sense that they notice or realize that they have this content. Thus, consciousness is different from awareness, the latter being stronger since if O is aware of X then O realizes or notices X. In contradistinction, O is conscious of X if O is in a state whose content is X. In the next section, I argue that the thoughts formed in late vision are contextual too, unlike thoughts in the space of reasons that I will call 'pure thoughts'.

3 Abduction in Late Vision

3.1 The Problem

Jackendoff (1989) distinguishes visual awareness from visual understanding. There is a qualitative difference between the experience of a 3D sketch and the experience of a 21/2D sketch. Visual awareness is awareness of Marr's 21/2D sketch, which is the viewer-centered representation of the visible surfaces of objects. One is also aware of the 3D sketch or of category based representations, however, this is not visual awareness but visual understanding, that is, seeing-as or seeing that as opposed to phenomenal seeing. The 3D sketch, which includes the unseen surfaces that are not represented in the 21/2D sketch, is a result of an inference. Jackendoff's views belong to the so-called belief-based account of amodal completion: the 3D sketch is the result of beliefs abductively inferred from the object's visible features and other background information from past experiences.

A more general problem is whether object identification that occurs in late vision (which is an abductive inference) should be thought of as a visual process properly speaking, that is, as involving visual awareness, or as case of discursive understanding involving inferences. If late vision involves conceptual contents and if the role of concepts and stored knowledge consists in providing some initial interpretation of the visual scene and in forming hypotheses about the identity of objects, one is tempted to say that this stage relies on inferences and, thus, differs in essence from the purely perceptual processes of early vision. It would be better to construe late vision as epistemic seeing.

I think that one should not assume either that late vision involves abductive inferences construed as inferential discursive state-transformations that constitutively involve thoughts in the capacity of premises in inferences whose conclusion is a recognitional belief, or that it consists in entertaining thoughts construed as the entities that figure in cognizing in the space of reasons, i.e., as pure thoughts. The reason is twofold. First, seeing an object is not the result of a discursive inference, that is, a movement in thought from some premises to a conclusion, even though it involves concepts and state transformations. Second, late vision is a stage in which conceptual modulation and perceptual processes form an inextricable link that differentiates late vision from the space of reasons even though late vision involves implicates beliefs that guide the formation of hypotheses, and an explicit belief of the form "that O is F" eventually arises in late vision. Late vision has an irreducible visual ingredient that makes it different from discursive understanding, and makes the thoughts that figure in late vision different from 'pure thoughts'.

Let me clarify two terminological issues. First, judgments are occurrent states, whereas beliefs are dispositional states. To judge that O is F is to predicate Fness to O while endorsing the predication (McDowell 1994). To believe that O is F is to be disposed to judge under the right circumstances that O is F. This is one sense in which beliefs are dispositional items. There is also a distinction between standing knowledge (information stored in LTM) and information that is activated in WM. The belief that O is F may be a standing information in LTM, a memory about O even though presently one does not have an occurrent thought about O. Beliefs need not be consciously or unconsciously apprehended, that is, activated in the mind, in order to be possessed by a subject; beliefs are dispositional rather than occurrent items. This is a second sense in which beliefs are dispositional. When this information is activated the thought that O is F is formed; all thoughts are occurrent states. I assume that beliefs are either pieces of standing information or thoughts that have not been endorsed and, thus, are not judgments. By "implicit belief" I mean the belief held by a person who is not aware that they are having this belief.

This paper examines whether the abductive processes that take place in late vision should be construed as inferences. My claim is that the processes in late vision are not discursive inferential processes, that is, processes that involve drawing propositions–conclusions from other propositions acting as premises either by applying (explicitly or implicitly) inferential rules. (These rules may also be represented, or may be realized by hardwired computational processors. In the latter case, the rules have no representational contents, or, if they do, these contents constitute tacit, nonconceptual knowledge.) These inferences are distinguished from "inferences" as understood by vision scientists according to whom any transformation of signals carrying information according to some rule is an inference. "Every system that makes an estimate about unobserved variables based on observed variables performs inference...We refer to such inference problems that involve choosing between distinct and mutually exclusive causal structures as causal inference" (Shams and Beierholm 2010).

There is another notion of inference that differs from the abovementioned because it is not restricted to making estimates based on observed variables only but makes estimates on the basis of object knowledge. Cavanagh (2011) argues that the processes that lead to the formation of a conscious percept constitute 'visual cognition' in virtue of using inferences. The construction of a percept is "the task of visual cognition and, in almost all cases, each construct is a choice among an infinity of possibilities, chosen based on likehood, bias, or a whim, but chosen by rejecting other valid competitors" (Cavanagh 2011, p. 1538). This process is an inference in that "it is not a guess. It is a rule-based extension from partial data to the most appropriate solution." (ibid. p. 1539). Thus, the selection process is an abductive inference.

According to Cavanagh (2011, p. 1545), the visual system does not rely on purely bottom-up analyses of the image that use only retinal information, such as sequences of filters that underlie facial recognition, or the cooperative networks that converge on the best descriptions of surfaces and contours. Instead, the visual system uses object knowledge, which is non-retinal, context-dependent information, and which is needed for the filling in that leads to the construction of the percept. This knowledge consists in rules that guide or constrain visual processing in order to solve various underdetermination problems; they provide the rule-based extension from partial data that constitutes an inference. These rules do not influence visual processing in a top-down way since they reside within the visual system; they are "from the side" (Gregory 2009).

There is indeed extensive evidence that there is a 'body of knowledge' that affects perception from within. The perceptual system does not function independently of any kind of internal restrictions. Visual processing at every level is constrained by as set of constraints that modulate information processing. Such constraints are needed because distal objects are underdetermined by the retinal image, and because the percept itself is underdetermined by the retinal image. Unless processing in the perceptual system is constrained by some 'assumptions' about the physical world, perception is not feasible. Most computational accounts hold that these constraints realize some reliable generalities of the physical world as it relates to the physical constitution and the needs of the perceiving agents. There is evidence that the physiological visual mechanisms reflect these constraints. Their physical making is such that they implement these constraints; the constraints are hardwired in perceptual systems.

These are Raftopoulos' (2009) 'operational constraints' and Burge's (2010) 'formation principles'. The operational constraints reflect higher-order physical regularities that govern the behavior of worldly objects and the geometry of the environment and which have been incorporated in the perceptual system through causal interaction with the environment over the evolution of the species. They allow us to lock onto medium size lumps of matter in our world, by providing the discriminatory capacities necessary for the individuation and tracking of objects. They enable perception to generate perceptual states that present worldly objects as cohesive, bounded, solid, and spatio-temporally continuous entities.

Among these operational constraints are: "local proximity" (adjacent elements are combined); "closure" (two edge-segments could be joined even though their contrasts differ because of illumination effects); "continuity" (the shapes of natural objects tend to vary smoothly and usually do not have abrupt discontinuities); "compatibility" (a pair of image elements are matched together if they are physically similar, since they originate from the same point of the surface of an object); and "figural continuity", (figural relationships are used to eliminate most alternative candidate matches between the two images). There is also a constraint to the effect that light falls on objects from above. These constraints clearly reflect regularities in the physical environment. Among the many different ways a state could be transformed to another state, only the transformations that accord with these regularities occur. Notice, though, that these regularities are not laws of nature. Usually light falls from above, but this is not always the case. Similar 'exceptions' apply to the other constraints. Thus, when states are transformed by means of computational processors that realize the constraints, the new state is the best fit to, or the best explanation of, the sensory data at hand, but it may not be the correct one; thus, the transformation is an abductive inference, in the wide sense of the term.

The constraints are not available for introspection, function outside the realm of consciousness, and cannot be attributed as acts to the perceiver. One does not believe implicitly or explicitly that an object moves in continuous paths or that it is rigid, though they use this information to parse objects. These constraints are not perceptually salient but one must be 'sensitive' to them if they are to be described as perceiving their world. The constraints constitute the *modus operandi* of the perceptual system and not a set of rules used by the perceptual system either as premises in inferences or as rules in inferences; the *modus operandi* of the visual system consists of operations determined by laws describable in terms of computation principles. They are reflected in the functioning of perception and can be used only by it, whereas "theoretical" constraints are available for a wide range of cognitive tasks. These constraints cannot be overridden since they are not under the perceiver's control; one cannot substitute them with another body of constraints even if they know that they lead to errors.

Haugeland (1998) argues that we share with non-concept possessing creatures various innate "object-constancy" and "object-tracking" mechanisms that automatically 'lock onto' medium sized lumps. These mechanisms provide the discriminatory capacities necessary for the individuation and recognition of objects in a bottom-up, nonconceptual way. Haugeland (1998, pp. 261–261) claims that the objective character of perception, that is the fact that perception is about objects *qua* objects, is due to the role of some normative standards that constitute thinghood. The "constitutive standards for thinghood" are cohesiveness and compatibility. These standards are in fact results of the operational constraints on perception that I discussed above.

Haugeland (1998, pp. 248–249) claims that neither the perceiver has a discursive cognizance of the standards in some explicit formulation, nor are these standards articulated as rules. Indeed, being hardwired, the constraints are not even contentful states of the perceptual system. It is arguable, thus, that neither the perceiver has a discursive cognizance of the standards in some explicit formulation, nor are these standards articulated as rules. Indeed, one could argue that being hardwired, the constraints are not even contentful states of the perceptual system, or, if they are, the contents are not conceptual, propositionally structured contents that could constitute some theory or other. Let me explain this. A neural state is formed through the spreading of activation and its modification as it passes through the synapses. The hard-wired constraints specify the processing, i.e., the transformation from one state to another, but they are not the result of this processing. They are computational principles that describe transitions between states in the perceptual system, and they constitute a computational processor. Although the states that are produced by means of these mathematical transformations have contents, there is no reason to suppose that the principles that specify the mathematical transformation operations are states of the system and, thus, are represented in the system; this is what the expression 'modus operandi; used above purports to convey. That is, even though the perceptual system by using the constraints operates in accord with the principles reflected in them, the perceiver does not represent the constraints.

What, then, about the claim that 'object knowledge' is needed for the filling in that it leads to the construction of the percept? If the operations that effectuate the filling-in are not represented in the system but are performed by hardwired computational processors, is it legitimate to talk about these processors realizing some object knowledge in the form of a set of rules concerning the physical environment and its geometry? This, as a matter of course, depends on what one is willing to count as knowledge.

We stated above that if the operational constraints are not states of the visual system but computational processors, they are not representations or beliefs of any form, either implicit or explicit. (Explicit beliefs are representations that are activated in persons, whereas implicit beliefs are representation that are stored in long-term memory but are not currently activated.) If the constraints are not states of the system, what is the status of the information they contain, that is, of the information included in the regularities about the environment and its geometry that the constraints realize? A first answer is that by not being states, the operational constraints do not have any contents; they are not semantic or mental entities of any kind. To think that they are is a mistake that, as Searle (1995) claims, cognitive scientists often commit. When they have an input and an output state both of which are mental states with representational contents, they tend to think that the processes that connect them are also mental states with some content. There is no reason, however, to assume this as the processes that connect the inputs with the outputs could be non-meaningful, that is, non-contentful, causal connections. According to this view, the function of the operational constraints in perception does not entail that perception is guided by 'object knowledge'.

Other philosophers think that such constraints are states of the system. They use the term 'tacit knowhow' to denote the information carried by states that are built into the system in a way that does not require that the states be represented in any form (Dennett 1983). This tacit knowhow is not represented anywhere in the system and is not a kind of knowledge because if it were we would have to say that birds, in the muscular system of which the laws of aerodynamics are hardwired, know aerodynamics.

There are philosophers who disagree with this view and think that hardwired computational processors realize in a system tacit *knowledge* of a particular set of rules or generalizations (Davis 1995, p. 329). "The rules would not have to be explicitly represented in any representational state of the system. Still less would knowledge of the rules be realized in a state of the same kind as an attitude state." Davis claims that tacit knowledge is not realized by states that are attitude states because tacit knowledge has two main characteristics. First, it is subdoxastic knowledge because tacit knowledge is not inferentially integrated with attitude states and it exists in special-purpose, separate sub-systems (Stich 1978). Second, with attitude states such as beliefs the concepts that are part of the semantic contents of the states must be concepts that are possessed by the believer; the believer should grasp the concepts of which the belief is constituted. This means that the beliefs have their representational contents conceptualized by the believer. The contents of tacit states, however, are not so conceptualized. Moreover, when a person is in a tacit state, they do not have, by being in that state, access to the contents of it.[2] In contradistinction, when a person is in a belief state, they entertain the content of the belief simply by being in that state. It follows that, according to the proponents of tacit knowledge, the operational constraints realize tacit, representational knowledge of the regularities of the physical environment and of its geometry. However, these are not conceptual representations.

Thus, irrespective of the interpretation one favors as to the status of the information realized by the operational constraints, that is, independent of whether one thinks that the constraints are merely causal connectors with no representational contents for the system, or whether this information constitutes some sort of tacit, non-representational knowhow, or whether it is some sort of tacit, representational knowledge, the operational constraints are not rules of inference that are looked-up

---

[2]The specification 'in virtue of being in that state' is needed to exclude cases where viewers form a belief about the content of one of their perceptual states not by being in that state but by drawing inferences from a background theory about perception. Suppose that the viewer is a vision scientist who knows about primal sketches and can form beliefs about her perceptual subdoxastic states whose contents implicate primal sketches. Even though the viewer entertains a belief about the contents of one of her perceptual states, this belief is the result of an inference from a body of theoretical knowledge about vision and not a direct consequence of the fact that the viewer is in a state with this particular content.

implicitly or explicitly by the visual system in order to perform its state to state transformations, or premises used in such transformations. Furthermore, the fact that perception relies on such constraints for successful function does not entail that perception is affected from concepts from within As we saw, in any interpretation of the information realized by the operational constraints, this is not conceptual content and, thus, it does not constitute some form of conceptual influence on perception from within. It follows that the existence of the operational constraints that are hardwired in perception does not entail that there is some sort of knowledge that determines perceptual processing. If theories are construed as carriers of knowledge about the world, the operational constraints by themselves do not entail the theory-ladenness of perception.

Thus, Burge (2010) is right to argue that the formation principles do not entail the theory-ladenness of perception:

> For many philosophers, the notion of computational states or explanations is theory-laden in a way that I do not intend. When I call states or explanations 'computational', I do not mean that there are transformations on syntactical items, whose syntactical or formal natures are independent of representational content [of the computed states]. I also do not mean that the principles governing transformation are instantiated in the psychology, or 'looked up', even implicitly in the system ... principles governing perceptual transformations ... are not the representational content of any states in the system, however unconscious. (Burge 2010, p. 95)

Spelke (1988, p. 458)—echoing Rock's (1983) view that the perceptual system combines inferentially information to form the percept (for example, from visual angle and distance information, one infers and perceives size)—argues "perceiving objects may be more akin to thinking about the physical world than to sensing the immediate environment." The reason is that the perceptual system, to solve the underdetermination problem of both the distal object from the retinal image and of the percept from the retinal image, employs a set of object principles and that reflect the geometry and the physics of our environment. Since the contents of these principles consist of concepts, and thus, the principles can be thought of as some form of knowledge about the world, perception engages in discursive, inferential processes. Against this, I argued above that the processes that constrain the operations of the visual system should not be construed as discursive inferences or as premises in inferences. They are hardwired in the perceptual circuits and are not conceptually represented in it. Thus, perceptual operations should not be construed as discursive inference rules. It follows that the abduction that takes place in late vision does not put perception on a par with thinking.

3.2 Late Vision, Hypothesis Testing, and Inference

I think that the states of late vision are not inferences from premises that include the contents of early vision states, even though it is usual to find claims that one infers that a tiger, for example, is present from the perceptual information retrieved from a visual scene. An inference relates some propositions in the form of premises with some other proposition, the conclusion. The objects and properties as they are represented in early vision, however, are not contents in the form of propositions, since they are part of the non-propositional, iconic content of perception. In late vision, the perceptual content is conceptualized but the conceptualization is not a kind of inference but rather the application of stored concepts to some input that enters the cognitive centers of the brain and activates concepts by matching their content. Thus, even though the states in late vision are formed through the synergy of bottom-up visual information and top-down conceptual influences they are not inferences from perceptual content.

Late vision, moreover, involves hypotheses regarding the identity of objects and their testing against the sensory information stored in iconic memory. One might think that inferences are involved since testing hypotheses is an inferential process even though it is not an inference from perceptual content to a recognitional thought. It is, rather, an argument of the form if A and B then (conclusion) C, where A and B are background assumptions and the hypothesis regarding the identity of an object respectively, and C is the set of visual features that the object is likely to have. A consists of implicit beliefs about the features of the hypothesized visual object. If the predicted visual features of C match those that are stored in iconic memory in the visual areas, then the hypothesis about the identity of the object is likely correct. The process ends when the best possible fit is achieved. However, the test basis or evidence against which these hypotheses are tested for a match, that is, the iconic information stored in the sensory visual areas, is not a set of propositions but patterns of neuronal activations whose content is non-propositional.

There is nothing inference-like in this matching. It is, instead, a comparison between the activations of neuronal assemblies that encode the visual features in the scene and the activations of the neuronal assemblies that are activated top-down from the hypotheses. If the same assemblies are activated then there is a match. If they are not, the hypothesis fails to pass the test. This can be done through purely associational processes of the sort employed in connectionist networks that process information according to rules and, thus, can be thought of as instantiating processing rules, without either representing these rules or operating on language-like symbolic representations. Such networks perform vector completion and function by satisfying soft constraints in order to produce the best output given the input and the task at hand.

In perceptual systems construed as neural networks, the fundamental representational unit is not some linguistic or linguistic-like entity but the activation pattern across a proprietary population of neurons. If one wishes to understand the workings of the brain, one should eschew sentences and propositions as the sole bearers of representations and meanings and reconceptualize representations as activation patterns. This means that the processing in the brain, that is, the transformation of the representational units to other representational units, is not the transformation of complex or simple symbols by means of a set of syntactic rules as in the algorithms that, according to the classical view, the brain is supposed to run. Instead, it is the algebraic transformation of activation patterns. The transformation is effected by the synaptic connections among the neurons as the signal passes from one layer to another.

Since discursive inferences are carried out through operations on symbolic structures, the processing in a connectionist network does not involve discursive inferences, although it can be described in terms of inference making. Thus, even though seeing an object in late vision involves the application of concepts that unify the appearances of the object and of its features under some category, it is not an inferential process. In the concluding section, I will discuss a ser of neural networks that simulate the formation of the percept with ambiguous figures and which not only respect but also explain the relevant empirical evidence.

I have said that the process that results in the formation of a recognitional thought/belief could be recast in the form of an argument from some premise to a conclusion. However, it is a non-discursive inference that does not involve the transition from a set of premises to a conclusion. For this reason, the formation of the perceptual recognitional thought is not an instance of reasoning from a set of premises that act as a reason for holding a thought to the thought itself. Admittedly, the perceiver can be asked on what grounds she holds the thought that O is F, in which case she may reply "because I saw it" or "I saw that O is F". However, this does not mean that the reason she cites as a justification of the thought is a premise from which she inferred the thought. The perceiver does not argue from her thought "I saw it to be thus and so" to the thought "It is thus and so". She just forms the thought on the basis of evidence included in her perceptual state in the non-inferential way described above. What warrants the recognitional thought "O is F" is not the thought held by the perceiver but the perceptual state that presents to her the world as such and such. "When one knows something to be so by virtue of seeing to be so, one's warrant for believing it to be so is that one sees it to be so, not one's believing that one sees it to be so." (McDowell 2011, p. 33).

To put it differently, if the state transitions in late vision were discursive inferences, a viewer should be able to justify the perceptual belief that late vision outputs by stating the reasons that led them to hold this belief; the whole process should be within the space of reasons. This is not true, however. In perception, the epistemic support or justification for perceptual beliefs accords with externalism and not internalism.[3] A viewer does not have to state the reasons on account of which they are justified in holding a perceptual belief. Instead, one's visual experience as if *X* is before them is a reason for believing that there is an *X* before them only because in one's world such a visual experience is *reliably* related to an *X* being before them.

[3]According to externalism, a belief is justified if the believer has formed that belief on the basis of a reliable belief formation process, which usually is outside the cognitive grasp of the believer. A belief is justified if "it comes from an epistemically, truth-conducively reliable process or faculty or intellectual virtue" (Sosa 2003, p. 109); perception and memory are examples of such intellectual virtues. This means that one can know that *X* is the case without having access to the reasons or processes that justify one's belief that *X* is the case; it suffices that this belief be produced in a causal reliable way by some process of belief formation.

3.3 Late Vision and Discursive Understanding

Even if I am right that seeing in late vision is not the result of a discursive abductive inference but the result of a pattern matching process that ensures the best fir with the available data, it is still arguable that late vision should be better construed as a stage of discursive understanding rather than as a visual stage. If object recognition involves forming a belief about class-membership, even if the belief is not the result of an inference, why not say that recognizing an object is an experience-based belief that is a case of understanding rather than vision.

One among the reasons why the beliefs formed in late vision are partly visual constructs and not pure thoughts is that the late stage of late vision in which explicit beliefs concerning object identity are formed constitutively involves visual circuits (that is, brain areas from LGN to IT in the ventral system). Pure thought, on the other hand, involves an amodal form of representation formed in higher centers of the brain, even though these amodal representations can trigger in a top-down manner the formation of mental images. The point is that amodal representations can be activated without a concomitant activation of the visual cortex. The representations in late vision, in contrast, are modal since they constitutively involve visual areas. Thus, what distinguishes late vision beliefs from pure thoughts is mostly the fact that the beliefs in late vision are formed through a synergy of bottom-up and top-down activation and their maintenance requires the active participation of the visual circuits. Pure thoughts can be activated and maintained in the absence of activation in visual circuits.

The constitutive reliance of late vision on the visual circuits suggests that late vision relies on the presence of the object of perception; it cannot cease to function as a perceptual demonstrative that refers to the object of perception (Burge 2010, p. 542). As such, late vision is constitutively context dependent since the demonstration of the perceptual particular is always context dependent. Thought, on the other hand, by its use of context independent symbols, is free of the particular perceptual context. Even though both recognitional beliefs in late vision and pure perceptual beliefs involve concepts, the concepts function differently in the two contexts.

> Perceptual belief makes use of the singular and attributive elements in perception. In perceptual belief, pure attribution is separated from, and supplements, attributive guidance of contextually purported reference to particulars. Correct conceptualization of a perceptual attributive involves taking over the perceptual attributive's range of applicability and making use of its (perceptual) mode of presentation. (Burge 2010, p. 545)

The singular and attributive elements in perception correspond to the perceived objects and their properties respectively. The attributive elements/properties guide the contextual reference to particulars/objects since the referent in a demonstrative perceptual reference is fixed through the properties of the referent as these properties are (re)presented in perception.

Concepts enter the game in their capacity as pure attributions that make use of the perceptual mode of presentation. Burge's claim that in perceptual beliefs pure attributions supplement attributions that are used for contextual reference to particulars may be read to mean that perceptual beliefs are hybrid states involving both visual elements (the contextual attributions used for determining reference to objects) and conceptualizations of these perceptual attributives in the form of pure attributions. In this case, the role of perceptual attributives is ineliminable. In late vision, unlike in pure beliefs, there can be no case of pure attribution, that is, of attribution of features in the absence of perceptually relevant particulars.

The inextricable link between thought and perception in late vision explains the essentially contextual, (Perry 2001; Stalnaker 2008, pp. 78–82) character of beliefs in late vision. The proposition expressed by the belief cannot be detached from the perceptual context in which it is believed and cannot be reduced to another belief in which some third person or objective content is substituted for the indexicals that figure in the thought (in the way one can substitute via Kaplan's characters the indexical terms with their referents and get the "objective" truth-evaluable content of the belief); the belief is tied to a idiosyncratic viewpoint by making use of the viewer's physical presence and occupation of a certain location in space and time; the context in which the indexical thought is believed is essential to the information conveyed.

4 Concluding Discussion: The Case of Ambiguous Figures: An Exemplification of the Matching Process

Ambiguous or bistable figures are usually two-dimensional figures that admit of two different organizations and depending on the organization the viewer can experience one of two mutually exclusive alternative percepts; one cannot see both figures at the same time, neither do they see an averaged figure, that is, a figure the results from an overlapping or weighting summation of the two alternative percepts. Research (Britz and Pitts 2011; Kornmeier et al. 2011; Pitts et al. 2007) shows that for each ambiguous figure there are some critical points focusing on which determines the percept through the role of spatial attention; that is, there are locations in the image fixation on which favors one or the other percept. Although these critical points are sufficient to cause the perception of a percept, they are not necessary since one can see one or the other percept even if one focuses on a neutral region in the image, and the simple introduction of a neutral fixation point does not stop reversals figure reversals.

The role of critical points as sufficient factors in determining the percept suggests that spatial attention can influence the way an ambiguous figure is perceived, a hypothesis that has received considerable experimental support (Long and Toppino 2004; Meng and Tong 2004; Toppino 2003). At this juncture an elucidation concerning the precise role of spatial attention is needed. Fixation points draw automatically spatial attention to a specific location in the figure, which in turn determines a certain organization of the image and, thus, the percept. Spatial attention is drawn to a specific location in the image by the biasing cue. When this happens, there is an increase in the base line firing rate of the neurons with receptive fields at the retinotopic position of the focus of spatial attention that prepares the neurons to process signals stemming from the focused areas. These shifts are independent of the stimulus; in fact, they are independent of whether a stimulus exists at the specific location (Murray 2008). These effects are called anticipatory effects and are established prior to viewing the stimulus.

Other research suggests that object/feature based attention can also influence the perception of ambiguous figures, (Leopold and Logothetis 1999; Long and Toppino 2004; Meng and Tong 2004; Toppino 2003). That is, the same preparatory activity via the increase of base line activity occurs with feature/object based attention. In this case, there is an increase of the base line firing rate of the neurons preferring the attended stimulus that the participant is instructed to attend to or for which a cue is presented before stimulus presentation. Both sorts of attention can override the bottom-up effects of fixation points when fixation conditions and the demands of intentional instructions are incompatible (for example, participants were instructed to maintain intentionally a designated orientation of the Necker cube), or when the figure is small enough not to allow for selective processing of different sets of focal features. This is done by causing covert attention shifts to other locations in the image where the selected features induce a different organization of the image (Britz and Pitts 2011; Pitts et al. 2007).

The anticipatory effects rely on top-down signals from parietal areas (in spatial attention) and frontal areas (in object/feature based attention) since these top-down signals effectuate the increase in the base line neuronal activity, although they do not affect the perceptual processing during stimulus viewing. Top-down parietal signals allocating spatial attention require 200–300 ms to bias the base line activity, while the top-down from frontal areas signals allocating feature attention require 300–500 ms, which means that voluntary top-down spatial and object/feature attention exert their respective effects at different time scales (Carrasco 2011; Liu et al. 2007).

The reversals when one focuses on a neutral point in the image, or the fact that one can choose one interpretation of the image over the other even when they fixate on a neutral point, are explained by the role of cognitively driven attention, whether it be spatial or feature/object based. The same attentional effects explain why one can shift from one interpretation to the other in the first place. It suffices to shift attention covertly from one critical location to another for the figure to reverse. Thus, the perception of ambiguous figures can be caused either by purely bottom-up factors or by a combination of both bottom-up, stimulus-driven effects and top-down, cognitive-driven effects.

Let us examine now whether the perception of ambiguous figures can be described in terms of the pattern matching processes of neural networks and, thus, to offer an alternative to the view that the interpretation of ambiguous figures is a sort of a discursive inference. The bistable figures can be modeled on the basis of the idea of dynamic bistability, according to which two stable states are formed in the phase space of the perceptual system.

To simulate the confluence of the top-down and bottom-up factors, Long and Toppino (2004) proposed a model that consists of interacting neural networks. Top-down effects from higher-order global processes affect representations in visual areas in IT and the extrastriate cortex (V4 and MT) and affect either the anticipatory activity before stimulus presentation or the recurrent processing during stimulus viewing. These areas also receive input from the feature-extraction level, presumably the early visual cortex. The locus of the resolution of the competing alternative interpretations of the ambiguous figure is at these intermediate levels (IT and V4, MT) because the representations of the alternative percepts of ambiguous figures occur in the primary and secondary visual cortex. Evidence suggests that IT neurons completely conform to the participants' experience, whereas neurons in V4 and MT are associated with the transition from one interpretation to the other (Andrews et al. 2002; Leopold and Logothetis 1999).

There are also models of the mechanism that may be responsible for the switch from the one state to the other. I consider here the mechanism of phase transition and metastability that is entailed by the property of intermittency. Such models can have metastable states and exhibit phase transitions that can be used to explain perceptual bistability and switching between alternative interpretation of the stimulus.

Such a neural network model was developed by van Leeuwen et al. (1997). The model is based on the fact that perceptual systems analyze external input for low and high spatial frequency features and gradually organize them in a globally coherent pattern, the percept. The salience of a feature determines the control parameter A for each neuron/node in the system. The more salient the input is to a receptive field (when, for, example, the input coincides with the preferred stimulus of the node the salience is maximal) the lower the value of A. Reducing A across a population of neurons the effect is the strengthening of the tendency for these neurons to couple and, thus, provide a stable percept. Amin, then, represents the figure in a visual scene, while Amax represents the background. Hence, synchronization of the system near Amin means that the representation of the figure becomes stable, while synchronization near Amax that represents the background is only temporarily. With ambiguous figures this means that the system synchronizes first in one stable interpretation, then it synchronizes temporarily in a non meta-stable state, and then spontaneously switches to the other meta-stable state, that is, the other alternative interpretation of the ambiguous figure. The equations governing the behavior of the system when A = Amax, and the fact that the system destabilizes switching between synchronized and unsynchronized states (between a stable percept to no percept and back to another percept) suggest that the interval between two switches varies stochastically in accordance with other evidence about switches of ambiguous figures (Meng and Tong 2004; Long and Toppino 2004).

It is important to note that the system goes into the circle of synchronized and unsynchronized activity driven by its internal dynamics and not by some external control. In other words, the self-organizing activity of the system can explain its behavior. As we have seen, in order for the system to be able to construct a percept, the synchronization must be within a range of parameter values. The automatic processes of the system can construct both local and global structures without the need for attentional processes to integrate local features into a coherent perceptual structure. This means that the perceptual system upon receiving input is able, driven by its own internal dynamics, to construct the percept without attentional control. However, the need to constrain synchronization entails that the system must operate within the range of an attentional parameter that controls the number of units that can become synchronized. This is a sort of non-specific attentional control whose role is not to reduce computational complexity and solve competition problems in terms of selecting neurons that encode one or the other feature, but to constrain the number of units that can become synchronized, so that no distorted patterns or no patterns at all are constructed.

Thus, the percept that prevails is determined by spatial attention, because spatial attention is the factor that constrains the number of neurons that are activated and can become synchronized by increasing the activation of the neurons whose receptive firs fall within the area of focus and by decreasing the activations of neurons with receptive fields a outside the focused area. This means that the features of the image at the focused location receive enhanced processing and determine the organization of the image and the ensuing percept.

Another simulation (Borisyuk et al. 2009) that uses phase transitions and meta-stability to model the perception of ambiguous figures posits a central unit surrounded by peripheral units. Each time an ambiguous figure is presented, depending on the initial viewing conditions the central unit co-opts, by partial synchronization, a group of peripheral oscillators and one interpretation is seen. Owing to its internal dynamics and the presence of noise in the sensory channels that renders the bursting activity and synchronization of the neurons irregular, this regime spontaneously switches so that the central unit becomes partially synchronized with another group of peripheral units and the alternative interpretation is seen. The presence of the central unit that selects sub-groups of peripheral modules allows the model to reflect the hierarchical organization of information processing in the brain. The central module models the parietal and frontal areas that are involved in the awareness and switching accompanying perception of ambiguous figures, while the peripheral nodes model the neuronal assemblies in visual areas. This way one can also model the effects of attention on perceiving ambiguous figures and it's effects on the temporal characteristics of the alternation perceptual process.

Work by Nakatani and Leeuwen (2006) may further elucidate the role of attention in perceiving ambiguous figures. In the previous paragraph, we saw that attention modulates perceptual switching through the modulatory effects of the central unit, which can be seen as the complex system of parietal and frontal areas involved in perceiving ambiguous figures and the concomitant perceptual switching. The research shows that when these areas participate in ambiguous figure perception, there is a synchronized activity in right parietal areas that are responsible for perceptual awareness and in the right frontal areas that are related to perceptual flexibility and, hence, to perceptual switching. Their research shows two cycles of synchrony in the gamma band; the first occurs 800–600 ms, and the second 400–200 ms before button pressing.

The same areas are also involved in top-down selective attention (Corbetta and Shulman 2002). The first period of synchronicity coincides with a drastic suppression of eye blinks that is related to attentional demands (Ito et al. 2003). The second period of synchronicity in the observed activity patterns in the fronto-parietal complex coincides with the maximum saccade frequency that reaches its peak at about 250 ms before the switch response. Since saccade frequency is associated with shifts of attention (Leopold and Logothetis 1999), the second period of synchronicity may reflect the final focus of spatial attention after a series of attentional shifts, which, as we have seen, by determining the critical points on the image that will be processed, also determines which interpretation of the ambiguous figure will be perceived.

Nakatani and Leeuwen (2006) explored the role of the activity of frontal and occipital cortex during switching episodes. They found that the theta activity in the frontal cortex is a general characteristic of the processing activity of viewers that perform frequent switches but is not specifically related to perceptual switching. The alpha band activity observed in the occipital cortex is related to frequent switches. Increased theta band activity in the frontal cortex is related to the concentration of attention to a task and to the inhibition of eye blink (Yamada 1998). Increased alpha activity in the occipital cortex is related to attention to the stimulus by enhancing the efficiency of information processing (Yamagishi et al. 2003). Thus, the frontal and occipital cortex activity during perceptual switches signifies the crucial role of attentional modulation of the perception of ambiguous figures and it's effects on the rate of perceptual switches.

In this section, I presented several connectionist models, all explaining and, thus supported by, considerable empirical evidence, to show that the percept formed in late vision can be the result of the pattern matching processes of neural networks, which are abductive ampliative inferences to the best fit, rather than the conclusion of abductive discursive inferences.

References

Andrews, T. J., Schluppeck, D., Homfray, D., Matthews, P., & Blakemore, C. (2002). Activity in the fusiform gyrus predicts conscious perception of Rubin's vase-face illusion. *Neuroimage*, *17*, 890–901.

Barr, M. (2009). The proactive brain: memory for predictions. *Philosophical Transactions of the Royal Society, London B Biological Sciences*, *364*, 1235–1243.

Block, N. (2007). Paradox and cross-purposes in recent work on consciousness. In N. Block (Ed.), *Collected Papers* (Vol. 1). Cambridge, MA: The MIT Press.

Borisyuk, R., Chik, D., & Kazanovich, Y. (2009). Visual perception of ambiguous figures: synchronization based models. *Biological Cybernetics*, *100*, 491–504.

Britz, J., & Pitts, M. (2011). Perceptual reversals during binocular rivalry: ERP components and their concomitant source differences. *Psychophysiology*, *48*, 1489–1498.

Bruner, J., & Goodman, C. (1947). Value and Need as Organizing Factors in Perception. *Journal of Abnormal and Social Psychology*, *42*, 33–44.

Burge, T. (2010). *Origins of objectivity*. Oxford: Clarendon Press.

Carrasco, M. (2011). Visual attention: the past 25 years. *Vision Research*, *51*, 1484–1525.

Cavanagh, P. (2011). Visual cognition. *Vision Research*, *51*, 1538–1551.

Corbetta, M., & Shulman, G. L. (2002). Control of goal-directed and stimulus-driven attention in the brain. *National Review of Neuroscience*, *3*, 201–215.

Davis, M. (1995). Tacit knowledge and subdoxastic states. In C. MacDonald & G. Macdonald (Eds.), *Philosophy of psychology: debates on psychological explanation*. Oxford: Blacwell.

Delmore, A., Rousselet, G. A., Mace, M. J.-M., & Fabre-Thorpe, M. (2004). Interaction of top-down and bottom up processing in the fast visual analysis of natural scenes. *Brain Research and Cognition*, *19*, 103–113.

Dennett, D. C. (1983). Styles of mental representation. *Proceedings of the Aristotelian Society*, *83*, 213–226.

Fabre-Thorpe, M., Delorme, A., Marlot, C., & Thorpe, S. (2001). A limit to the speed of processing in ultrarapid visual categorization of novel natural scenes. *Journal of Cognitive Neuroscience*, *13*, 171–180.

Gregory, R. L. (2009). *Seeing through illusions*. Oxford, UK: Oxford University Press.

Haugeland, J. (1998). *Having thought*. Cambridge: Harvard University Press.

Ito, J., Nikolaev, A. R., Luman, M., Aukes, M. F., Nakatani, C., & Leeuwen Van C. (2003). Perceptual switching, eye movements, and the bus paradox. *Perception*, *32*(6), 681–698.

Jackendoff, R. (1989). *Consciousness and the computational mind*. Cambridge, MA: MIT Press.

Johnson, J. S., & Olshausen, B. A. (2005). The earliest EEG signatures of object recognition in a cued target task are postsensory. *Journal of Vision*, *5*, 299–312.

Kihara, K., & Takeda, Y. (2010). Time course of the integration of spatial frequency-based information in natural scenes. *Vision Research*, *50*, 2158–2162.

Kornmeier, J., Pfaffle, M., & Bach, M. (2011). Necker-cube: Stimulus-related (low-level) and percept-related (high-level) EEG signatures early in occipital cortex. *Journal of Vision*, *11*(9), 1–11.

Kosslyn, S. M. (1994). *Image and brain*. Cambridge, MA: MIT Press.

Leopold, D. A., & Logothetis, N. K. (1999). Multistable phenomena: Changing views in perception. *Trends in Cognitive Science*, *3*(7), 254–264.

Liu, T., Stevens, S. T., & Carrasco, M. (2007). Comparing the time course and efficacy of spatial and feature-based attention. *Vision Research*, *47*, 108–113.

Long, G. M., & Toppino, C Th. (2004). Enduring interest in perceptual ambiguity: Alternating views of reversible figures. *Psychological Bulletin*, *130*(5), 748–768.

McDowell, J. (1994). *Mind and world*. Cambridge, MA: Harvard University Press.

McDowell, J. (2011). *Reception as a Capacity for Knowledge*. Milwaukee, Wi: Marquette University Press.

Meng, M., & Tong, F. (2004). Can attention selectively bias bistable perception? *Journal of Vision*, *4*, 539–551.

Murray, S. O. (2008). The effects of spatial attention in early human early visual cortex are stimulus independent. *Journal of Vision*, *8*(10), 1–11.

Nakatani, H., & van Leeuwen, C. (2006). Transient synchrony of distant brain areas and perceptual switching in ambiguous figures. *Biological Cybernetics*, *94*, 445–457.

Perry, J. (2001). *Knowledge, possibility, and consciousness*. Cambridge, MA: MIT Press.

Pitts, M., Nerger, J., & Davis, T. J. R. (2007). Electrophysiological correlates of perceptual reversals for three different types of multistable images. *Journal of Vision*, *7*(1), 1–14.

Raftopoulos, A. (2013). The cognitive impenetrability of the content of early vision is a necessary and sufficient condition for purely nonconceptual content. *Philosophical Psychology*, 1–20. doi:10.1080/09515089.2012.729486.

Raftopoulos, A. (2009). *Cognition and perception: How do psychology and the neural sciences inform philosophy?*. Cambridge, MA: MIT Press.

Rock, I. (1983). *The logic of perception*. Cambridge, MA: MIT Press.

Searle, J. R. (1995). Consciousness, explanatory inversion and cognitive science. In C. MacDonald & G. Macdonald (Eds.), *Philosophy of psychology: Debates on psychological explanation*. Oxford: Blacwell.

Shams, L., & Beierholm, U. R. (2010). Causal inference in perception. *Trends in Cognitive Science*, *14*, 425–432.

Sosa, E. (2003). Beyond internal foundations to external virtues. In L. Bonjour & E. Sosa (Eds.), *Epistemic justification: Internalism vs. externalism, foundations vs. virtues*. Oxford: Blackwell Publishing.

Spelke, E. S. (1988). Object perception. In A. I. Goldman (Ed.), *Readings in philosophy and cognitive science* (pp. 447–461). Cambridge, MA: MIT Press.

Stalnaker, R. C. (2008). *Our knowledge of the internal world*. Oxford: Clarendon Press.

Stich, S. (1978). Beliefs and subdoxastic states. *Philosophy of Science*, *45*, 499–518.

Toppino, T. (2003). Reversible-figure perception: mechanisms of intentional control. *Perception and Psychophysics*, *65*, 1285–1295.

Treisman, A. (2006). How the deployment of attention determines what we see. *Visual Cognition*, *14*, 411–443.

Tye. M. (2009). *Consciousness revisited: Materialism without phenomenal concepts*. Cambridge, MA: The MIT Press.

van Leeuwen, C., Steyvers, M. & Nooter, M. (1997). Stability and intermittency in large-scale coupled oscillator models for perceptual segmentation. *Journal of Mathematical Psychology*, *41*, 319–44.

Yamada, F. (1998). Frontal midline theta rhythm and eye linking activity during a VDT task and a video game. *Ergonomics*, *41*, 678–688.

Yamagishi, N., Callan, D. E., Goda, N., Anderson, S. J., Yoshida, Y., & Kawato, M. (2003). Attentional modulation of oscillatory activity in human visual cortex. *Neurotic Age*, *20*, 98–113.

The Correspondence Principle, Formal Analogy, and Scientific Rationality

Jeongmin Lee

Abstract This paper offers a case study in philosophical history of quantum theory, focusing on the role of Bohr's correspondence principle in the creation of the new mechanics. I argue that the principle is best understood as formal or symbolic analogy in the strictly Kantian sense of analogy. By showing how new quantum formalism embodies this philosophically loaded principle, I claim that the emergence of the new mechanics is unintelligible unless we take into account Bohr's Kantian philosophy before 1925. This may shed fresh light on the problem of scientific rationality.

> In so far as the philosophy of science draws on the history of science, and forms its picture of the *bona fides* and rationality of science on that basis, *then if the history is seriously misguided, so too will be the philosophy 'modelled' on it.* (Sharrock and Read 2002, p. 11, original emphasis).

1 Introduction

There is little doubt that Bohr's correspondence principle played a crucial role in the development of quantum theory, but the nature of the principle is still much in dispute. Bohr was certainly aware that the new formulation of quantum mechanics by Heisenberg and others utilized and extended the full potential of the principle. For instance, right after the publication of Heisenberg's (1925) paper, Bohr asserts that the "whole apparatus of the quantum mechanics can be regarded as a precise formulation of the tendencies embodied in the correspondence principle" (Bohr 1972, 5.280). Far from being Bohr's idiosyncratic assertion, a similar expression appears almost verbatim in the introduction to the *Dreimännerarbeit* of Born, Heisenberg, and Jordan: the new theory, according to them, "can itself be regarded as an exact formulation of Bohr's correspondence considerations (*Korrespondenzgedankens*)" (Born et al. 1926, p. 322). Both Bohr and the young generation of physicists recognized the central importance of the principle for the creation of the new mechanics,

J. Lee (✉)
Department of Philosophy, University of Seoul, Seoul, Korea
e-mail: philist@hotmail.com

L. Magnani et al. (eds.), *Philosophy and Cognitive Science II*,
Studies in Applied Philosophy, Epistemology and Rational Ethics 20,
DOI 10.1007/978-3-319-18479-1_10

but the meaning of "tendencies" or "considerations" associated with the principle is what is open to scholarly dispute.

Most textbook and historical accounts of the principle describe it as a requirement for an asymptotic agreement between the two theories, classical and quantum mechanics. According to this interpretation, the principle is a requirement that the new mechanics should reproduce correct classical results in the limit where many quanta are involved or the quantum number gets large. There is an obvious parallel with the theory of relativity, where various relativistic formulas approximate to the standard classical ones in the limit of low velocities. Understood in this way, the correspondence principle is a general requirement for a successor theory in physics. The new theory should preserve the results of the previous theory in the domain where the latter makes successful predictions about observable phenomena. The "correspondence" in question is then supposed to hold between two physical theories, and once it holds, a classical formula becomes a limiting case of a corresponding quantum formula.

This customary understanding of Bohr's principle, however, is not a historically accurate picture of it and leaves the above claims of quantum physicists unintelligible. In fact, recent historical studies put a lie to this almost universal misunderstanding of Bohr's original principle. As a pedagogical reconstruction, it may work fine for students of physics, but it is a serious misconception as applied to the early history of quantum theory. At least, it is not how Bohr himself understood the principle. When he was later asked about this point by Rosenfeld, his long-time collaborator and the early editor of his *Collected Works*, Bohr was quite dismissive of this type of misunderstanding: "the requirement that the quantum theory should go over to the classical description" in the classical limit is "not the correspondence argument ... not at all a principle. It is an obvious requirement for the theory" (Rosenfeld 1973, p. 253). Rosenfeld in the same article also records Bohr's colloquial preference for the correspondence "argument" instead of "principle," and this record suggests something significant. For Bohr and his followers, the correspondence principle was primarily a certain type of "considerations", or "tendencies" of thought, rather than a formal principle that can be uniquely cashed out in terms of precise mathematical formulas. Probably this feature is one of the reasons that the principle has often raised the suspicion that there is no exact formulation of the principle at all.

I am not going to spend a lot of space trying to dispel these common misconceptions, however. Instead, I will propose an alternative interpretation of the principle as *formal analogy* and relate my reading to the birth of quantum mechanics for additional support. It is exactly in this analogical sense that both Bohr and quantum physicists understood the principle, and without such an understanding, the entire formulation of new quantum mechanics would have been inconceivable. Thus there is a certain historical necessity that quantum theory could develop in no other way than the one envisaged by Bohr. Moreover, as I will show, Bohr's own conception of formal analogy is uniquely Kantian in that far from being an isolated usage, it brings in a whole cluster of related Kantian notions, such as pictures, intuitions, symbols, and (classical) concepts.[1] After discussing these issues, I conclude with a reflection on how this revised history can throw fresh light on by now a worn-out debate on scientific rationality.

2 The Correspondence Principle as Formal Analogy

We can trace the germ of the correspondence principle back to Bohr's 1913 trilogy on the hydrogen atom. Sending a revised copy of his paper to Rutherford, Bohr speaks of "the most beautiful analogi [sic] between the old dynamics and the considerations" used in the original paper (2.584). It will turn out that the analogy here is essentially of the same nature as the correspondence principle he formulated around 1920. Therefore, it will be useful to look into the original form of the analogy first.

In his 1913 model of the hydrogen atom, Bohr assumes that a single electron occupies one of the stationary states. When the electron makes a discontinuous state transition, it emits electromagnetic radiation with the frequency proportional to the energy difference between the two levels. When the quantum number N involved in the transition is large, the energy levels are almost equidistant, and the frequencies emitted are almost integer multiples of the frequency of transition between the adjacent levels, e.g., from N to $N-1$. The quantum frequency ω of radiation in that region can be approximated by the formula

$$\nu(N \rightarrow N-n) = n\omega \tag{1}$$

Now in classical electrodynamics, the position x of an electron can be expressed in terms of the Fourier series, i.e., the sum of the Fourier components, whose frequencies (boldfaced below) are integer multiples of the fundamental frequency ω:

$$\begin{aligned} x(t) &= \sum_{\tau} C_{\tau} cos 2\pi(\tau\omega t + c_{\tau}) \\ &= C_1 cos 2\pi(\boldsymbol{\omega} t + c_1) + C_2 cos 2\pi(\mathbf{2}\boldsymbol{\omega} t + c_2) + C_3 cos 2\pi(\mathbf{3}\boldsymbol{\omega} t + c_3) + \cdots \end{aligned} \tag{2}$$

In the classical formula (2), Fourier components with frequencies 2ω, 3ω, ... are all coexistent with the component with the fundamental frequency ω. Still, according to Bohr, the quantum formula (1), whose emitted frequency is one of the ω, 2ω, 3ω, ... can be understood in "analogy with the ordinary electrodynamics" represented in the formula (2) (2.174). The possible transitions between two quantum states in (1) "correspond" in a certain sense to the various harmonic components in the motion of the classical system, and this general connection is what he later named the correspondence principle.

[1] It is Chevalley (1995) who first described Bohr's Kantian connection in terms of "cluster concepts."

However, as Bohr quickly points out in his ensuing publications, there are obvious limitations to the analogy developed so far. First, there are physical differences between the two theories of electrons. Especially in quantum theory, there is as yet no well-defined physical picture of "how and why the radiation is emitted" (2.295). In classical electrodynamics, the continuous motion of an electron can be resolved into the Fourier components, each of which oscillates with a certain frequency, thereby emitting radiation of the same frequency. There is a mechanical explanation as to the cause of the radiation, grounded in the spatial motion of an electron. In quantum theory, on the other hand, we have very little clue about the motion of an electron inside the atom. Even though stationary states of an electron in the old quantum theory are usually described in terms of various circular or elliptic orbits, they cannot be conceived literally as the motion of the electron. For the simple yet mysterious reason that the quantum transition is discontinuous, electrons do not pass through the intervening space between stationary states. As Bohr suggests, we cannot "picture the mechanism of transition" (3.48) and should not forget that a "quantum leap" or "quantum jump" is just a convenient way of covering up that unknown mechanism. In other words, due to the essential discontinuity at the foundation of quantum theory, there is no definite mechanism of radiation comparable to the classical one. The state transition in Bohr's model provides no causal explanation for the radiation.

From the earliest days of his scientific career, Bohr recognized the fundamental and unbridgeable gap between the underlying physics of the two theories and the lack of mechanical explanations in the quantum theory. There is "no question of a gradual approach between the character of the radiation process" on the two theories. There is no way you can graft the classical theory into the quantum realm because the validity of the classical results at large N is only a "concealment of the difference in principle" between as yet unknown "actual [quantum] mechanism" and the "continuous laws of classical conceptions" (3.375, 468). The analogy discussed above provides little indication as to the nature of transition between stationary states. We cannot infer the quantum mechanism from the classical one based on some vague similarities. Rather, as Bohr realized from the beginning, the former should be constructed from the ground up only by the wholesale replacement of the classical mechanism.

The second limitation of the original analogy concerns numerical discrepancies at small quantum numbers. The expression for the frequency of radiation, $\nu(N \rightarrow N-n) = n\omega$, is not applicable when N is even moderately small. E.g., the frequency of the green line $H_\beta(N = 4 \rightarrow 2$, 617 THz) is by no means twice as high as that of the red line $H_\alpha(N = 3 \rightarrow 2$, 457 THz). Bohr's analogy is especially vulnerable to criticism on this account, and some critics have gone so far as to claim that the correspondence principle in this form is no principle at all, but some kind of ad hoc device to cover up deficiencies of the old quantum theory. More puzzling is Bohr's own remark that the analogy holds as a matter of a general principle without any restriction on the quantum numbers in question (3.249). How are we to make sense of Bohr's remark despite the numerical discrepancies already exposed at this point?

The first thing to notice is that numerical discrepancies in question did not actually matter much to Bohr. For him, the gist of the correspondence principle was not to gain a quantitative approximation at small quantum numbers by extrapolation from the results at large N. Despite the fundamental physical differences and numerical discrepancies at small N, the analogy "must have a deeper significance" according to Bohr. Especially interesting in this regard is the following characterization of the correspondence principle with the above example. The green line H_β may be considered in a certain sense to be an 'octave' of the red line H_α and "this [octave] relationship we call the 'correspondence principle'" (3.249-250; 4.348). Now the key question is in what sense we can regard the two lines as related by an octave.

Certainly, it cannot be in a numeral sense if an octave means doubling the frequency. We should keep in mind that the term "the correspondence principle" was introduced only in 1920, and the above locution is one of its earliest instances. In the meantime, the principle was known as "(formal) analogy" in Bohr and other physicists's writings. For example, the first edition of Sommerfeld's *Atombau* (1919), the bible of the old quantum theory, has *Analogieprinzip* instead of *Korrespondenzprinzip* found in the later editions (Tanona 2002, p. 77). In his letter to Bohr, Sommerfeld even refers to it as the "formal principle of analogy" (*formales Analogie-Princip*) (3.688). In other places, Bohr specifically mentions the correspondence "in the formal sense."[2] If we take all these circumstances into account, a reasonable conclusion should be that the correspondence principle is the octave relation in the formal-analogical sense. Now the next question is how to understand the formal sense of analogy.

In this connection, many scholars have directed attention to Kant's notion of analogy as one possible source of Bohr's correspondence principle (e.g., Chevalley 1994; Lee 2006; Pringe 2007). The analogy here refers to the one found both in the *Prolegomena* and the third *Critique*, not the "analogy of experience" in the first *Critique*. Analogy, according to Kant,

> does not signify (as is commonly understood) an imperfect similarity of two things, but a perfect similarity of relations between two quite dissimilar things (*Prolegomena*, §58).

Høffding, Bohr's philosophical mentor, took up exactly this notion of analogy in his *Mind* article: "analogy is likeness of the relations of different objects, not likeness of single qualities" (Høffding 1905, p. 200). Indeed in 1922, during the heyday of the correspondence principle, Bohr had a meeting with Høffding and discussed the role of analogy in the natural sciences (10.513-4).

What is unique about Kant's notion of analogy is that he takes it to be a similarity of relations, not of objects or qualities. If A is analogous to B in this sense, A and B themselves should be relations. Moreover, A and B may concern quite dissimilar objects, but the respective relations should bear a perfect similarity. As an illustration, consider the analogy between a heart and a water pump. The common understanding is that the analogy holds for these two objects since they are similar in certain aspects, but not in other aspects. What Kant says is that this is a wrong way of viewing an analogy. Rather, an analogy holds for two relations, here between the relation of the heart to blood (say H:B), and the relation of the pump to water (P:W). The heart

---

[2]Compare two similarly phrased sentences in (3.48) and (3.246).

and the pump (and blood and water) are two dissimilar things. Still, the relation H bears to B is perfectly similar to the relation P bears to W. In order for this perfect similarity to hold, every other quality of the objects concerned should be abstracted except for the bare-bones relation, "pumping".

Translated into Bohr's language, "correspondence" in question does not hold between two objects or even theories. Instead, two formal relations in respective theories are in correspondence with each other. In the classical theory, the relation between the fundamental frequency and the second harmonic is an octave, which is also numerically 1:2. In the quantum theory, the "formal" (not numerical) relation of H_α to H_β is also 1:2 in the sense that each corresponds to the jump of 1 ($N = 3 \rightarrow 2$) and jump of 2 ($N = 4 \rightarrow 2$), respectively. Electrons as classical oscillators are quite dissimilar from electrons in stationary states, and so are the underlying mechanisms of radiation. The quantum mechanism of transition is not even known. Notwithstanding the difference in the underlying physics, the spectrum (H_α and H_β) reflects the orbital motion (jumps of 1 and 2) in quantum theory in "exactly the same way as in the classical theory" (4.149). The perfect similarity holds because the relation of H_α to H_β is formally, but not numerically, identical to the relation of the fundamental to the second harmonic (octave) in the classical theory. The analogy is entirely general and applicable to other formal relations in quantum theory, some of which were yet to be discovered by 1922 (see Table 1). The generality of the formal analogy in the quantum domain is why Bohr often said that the correspondence principle is a principle or a law of quantum theory even though it was obtained from classical analogy and still lacked precise mathematical contents.

At this point, it may be objected that the Kantian connection of Bohr's principle still remains tenuous. Before answering this objection more fully in the next section, I'd like to point out that neglecting the uniquely Kantian notion of analogy is what led some scholars to misconstrue the whole point of the correspondence principle. For example, Tanona (2002, p. 9) dismisses the analogical aspect altogether on the ground that treating the correspondence principle as a "broader metaphor or formal analogy" does not address "Bohr's actual views of the role" of the principle. Well, formal analogy *is* the correspondence principle. The mistaken view of an analogy as

Table 1 The correspondence principle as formal analogy

	A	B
Formal analogy	Classical harmonic components (C)	Quantum transitions (Q)
Frequency of radiation	$\tau\omega$	$\nu(N \rightarrow N - n) = n\omega$ (high n)
Intensity of radiation	$\lvert C_\tau \rvert^2$ (amplitude)	$A^n_{n-\tau}$ (transition probability)
Position	$x(t) = \sum_\tau C_\tau cos 2\pi(\tau\omega t + c_\tau)$	?
Governing equations	Hamilton's equations of motion	?
Picture-mechanism	Classical motion (space-time, causal)	?

The analogy is: as C is to Q, so is A to B for other relations (not necessarily numerical ratios)

an imperfect similarity of two things is what is behind Tanona's apposition "broader metaphor or formal analogy." Darrigol (1992, p. 171), who treats extensively the role of analogies in both classical and quantum theories, also misses the Kantian overtone of Bohr's analogy and instead discusses the deductive and inductive aspects or uses of the correspondence principle. This idea was also taken up by Tanona, who characterizes the principle as a "deduction from phenomena" or an "epistemological bridge principle" that connects the spectral phenomena with the atomic models (Tanona 2002, pp. 8, 6).

While not denying that these ideas are some implications of the principle, I think they come up short in touching the nature or the "deeper significance" of the principle Bohr had in mind. At least, they are not how Bohr formulates his principle. First of all, in the application of the correspondence principle as formal analogy, more than two terms are involved. The relata of the "correspondence" are neither phenomena and theory, nor classical and quantum theories per se, but two formal relations in the respective theories, each of those concern two terms. Also, I think to associate logical terminology or empirical phenomena with the principle is potentially misleading. The analogy is a "question of a purely quantum theory theorem" (3.178) and is made to stay within the theoretical discourse. This theoretical concern of Bohr, along with the Kantian aspect and the formal character of the correspondence principle, will become evident as we move on to the next phase of Bohr's thought.

3 Symbolic Analogy and the Birth of Quantum Mechanics

Around 1922, Bohr's thought underwent a very significant development, which has hitherto been unnoticed by other scholars. The key idea here can be put as the transition from formal to symbolic analogy, the latter of which would eventually bring about the quantum revolution three years later. This idea was made available by eliciting the symbolic dimension latent in the correspondence principle, at the expense of space-time pictures inside the atom. Now the meaning of symbols and pictures must be explicated in the context of Kantian philosophy. Once done, the philosophical significance of the quantum revolution will emerge.

As indicated above, the point of the correspondence principle is not to obtain a numerical approximation at small N. The correspondence principle was intended to throw light on formal relations or the entire mechanisms in quantum theory, which were still in the dark. Just in such a case, the correspondence principle can provide some clue as to the correct formalism and mechanism, in light of the analogous situations in the classical theory. The eventual purpose of the correspondence principle, as Bohr once puts it, is to devise "quantum kinetics" (3.469) in analogy with classical equations of motion. Once we have correct formalism in place, again by analogy, the quantum mechanism might be inferred from it.

From his 1913 model of the hydrogen atom to its various modifications, it became increasingly obvious that the classical picture of the electron breaks down in the atomic domain. Here the concept of picture must be understood not in its literal sense but as it was understood by Bohr's contemporary physicists as a theoretical representation in general.[3] For example, the Fourier representation of a moving electron in the classical electrodynamics is a kind of picture even though the electron is not literally oscillating with all possible frequencies. Rather, the motion of an electron can be decomposed theoretically by the Fourier analysis. It is in this sense of "re-presenting" a physical system and its state of motion in a mathematical space that such a picture is said to be "visualizable."[4] When Bohr sometimes refers to the classical space-time description as an "intuitive picture," it is this feature which justifies such usage.

Now the key question about various atomic models was to what extent the classical picture should be limited, or even whether it should be completely renounced. Initially, Bohr was cautiously optimistic that a picture of atomic processes may be developed in analogy with classical ideas (3.366/402; 3.374/413). Again, the analogy is only formal, bearing no obvious similarities to the old picture. The quantum mechanism underlying the new picture would require "radical departure" from the classical conception. However, beginning in 1922, even this modest optimism about the possibility of a picture gave way to a more skeptical attitude toward it. The complete theory of the atom in the future might be so unique a physical theory that the quantum picture or mechanism is in principle unobtainable. Instead, we may as well develop quantum formalism to the exclusion of a space-time picture. These are conceptual possibilities Bohr starts to entertain around this period, which would become fully real before long.

Another indication that Bohr became particularly sensitive to these issues is the frequent appearance of the term "symbols", which had been absent up to this point in Bohr's writings. Symbols or "mechanical symbols" refer to quantities that are originally defined by a classical mechanism but used instead for the formulation of quantum relations. For example, the "symbolic" quantum condition for the angular momentum is given by $J = nh$, which fixes the stationary states at integer multiples of h. What is symbolic about such quantities is that they are now dissociated from the classical picture of orbits or oscillators and relocated in the purely formal context of the quantum theory. Nothing is quite "rotating," or is represented as rotating, in its classical sense here. Nor is there any hint of quantum mechanism to be discovered. "The interpretation of atomic phenomena," says Bohr, "does not involve a description of the mechanism of the discontinuous processes" (5.101). Concepts in atomic theory are formal in the sense that "they do not provide a visual picture of the sort" provided by the classical theory (4.482). With the symbolic dimension added to the formal analogy, what is formal or symbolic is explicitly contrasted with intuitive pictures.

---

[3]See Hertz (1956) [1894] for the "classical" conception of picture.

[4]German versions, if available, always have "*anschaulich*" in this place, which takes a distinctively Kantian overtone. Unfortunately, there is no uniform rendering of this key concept of Bohr and other quantum physicists in English and has been variously translated as intuitive, illustrative, evident, visualizable, etc. For example, Heisenberg's celebrated 1927 paper on the uncertainty relation is titled "*Über den anschaulichen Inhalt der quantentheoretischen Kinematik und Mechanik*," which was put into "physical' content of ..." in Wheeler and Zurek (1983, p. 62). "Intuitive pictures" translates "*anschaulicher Bilder.*"

Again we can trace this unique notion of symbol and analogy back to Kant. In the earlier example of the heart and the pump, both objects were sensible, and empirical intuition could be given for them. However, if one of the poles in the analogy becomes an idea of reason like God for which no intuition can be given, the analogy becomes merely symbolic. Take the analogy between the relation of parents to their children on the one hand, and the relation of God to mankind on the other. We have no idea about the latter except that there is a structural identity between the two relations, parental love of children and divine benevolence toward mankind. The analogy holds not in the least because of a partial similarity human love bears to divine benevolence, but because of the perfect similarity between them. We can transfer the formal, but not sensible, aspect of the mundane relation to an entirely different context of the divine relation, and the former as an object of intuition then becomes a symbol for the latter as a supersensible thing. This unique notion of symbolic analogy, as contrasted with the intuitive mode of presentation, was soon to be embodied in the correspondence principle, as we will see.

What might appear paradoxical in this line of development is the BKS paper of 1924. In the BKS theory, each electron is associated with a set of virtual oscillators with various frequencies, which coincide with the frequencies of transitions the electrons can undergo. The electromagnetic field is produced by the virtual motion of these oscillators, and electrons interact and make transitions through the mediation of such virtual field. By this maneuver, Bohr tried to save a "*picture* as regards the time-spatial occurrence of the various transition processes" (5.107, my emphasis). In the light of our above discussions, however, the BKS theory was an aberration, a kind of Bohr's philosophical experiments rather than a serious attempt to formulate a final theory. When Bothe and Geiger's experiment finally disconfirmed the BKS theory early in 1925, Bohr finally accepted the ultimate breakdown of a space-time description of the atom. On a conceptual level, however, Bohr was "quite prepared"; the BKS theory was "more an expression of an endeavour . . . than a complete theory" (5.79). The problem, finally crystallized in Bohr's mind, was with the unlimited applicability of "intuitive pictures" (*anschaulicher Bilder*), the renunciation of which is "characteristic of the formal treatment of problems of radiation theory" (5.190-3/206-8).

Heisenberg acknowledged that the BKS paper "was very helpful in leading him in the right direction" (Rosenfeld 1973, p. 258) i.e., away from intuitive pictures. What that "right direction" might be is suggested by another testimony of Heisenberg about the same period. In a long and heated discussion between Bohr and Kramers in 1924, even before the breakdown of the BKS theory, "the necessity for detachment from the intuitive models was for the first time stated emphatically and declared to be the guiding principle in all future work" (Rozental 1967, p. 98). Probably Heisenberg was exposed to the idea for the first time, but Bohr had entertained such a conceptual possibility for two years by then.

Throughout this period, the correspondence principle as a clue to quantum formalism has not been discarded though the emphasis has been shifted. Earlier, the principle stood for formal analogies, which were not necessarily exclusive of the corresponding pictures. It was a perfect similarity of relations between two objects, one of which was classical and the other was quantum-theoretical. However, correspondence as a symbolic analogy implies exactly such renunciation of space-time pictures in the quantum domain. The principle now concerns merely symbolic transference of formal relations into the quantum theory. When new formalism was invented in accordance with the guidelines offered by the principle, quantum theory was born. Bohr's dictum that the correspondence principle is a principle of quantum theory, old and new alike, was finally vindicated.

Indeed, concurrent with the rise and fall of the BKS model in 1924–25, there appeared a series of papers in which symbolic translations with various classical formulas were carried out more or less systematically, guided by the correspondence arguments. Now the correspondence principle was understood by Heisenberg, Born, and others exclusively as the transference or transcription of the classical formulas into the quantum language, detached from the intuitive models. This program, "sharpening of the correspondence principle" as Heisenberg called it, turned out to be the royal road to matrix mechanics.

The first step toward this direction was taken by Kramers in his short paper on the dispersion phenomena. In this paper, Kramers succeeded in obtaining the first-ever quantum formula for dispersion by removing classical concepts concerned with physical pictures (orbits). He took the classical dispersion formula and replaced the oscillation frequency and the Fourier amplitude in it with a radiation frequency and a transition probability of quantum theory. The quantum quantities depend on two discrete values and "allow of a direct physical interpretation," claims Kramers. The classical quantities, by contrast, must be understood through the classical picture, i.e., the motion of periodic systems (5.44-5). This contrast between physically meaningful quantities and superfluous classical pictures soon became a common thread running through the "sharpening" program.

In the meantime, Born (1924) developed more systematically Kramers' ideas into the perturbation theory of coupling. The crucial step in Born's treatment was to import the classical perturbation formulas and to provide necessary quantum-mechanical translations of them. Such "quantization procedure" involves replacing continuous physical quantities, e.g., the classical $\tau\omega$, with corresponding "transition quantities" depending on two stationary states, $\nu(n, n \pm \tau)$ in this case. Born once called this process the "formal passage from classical mechanics to 'quantum mechanics' (*formalen Ubergang von der klassischen Mechanik zu einer "Quantenmechanik"*)," where the latter word appears for the first time.

Apart from its historic interest, noteworthy is physicists' original conception of the successor theory suggested by the phrase. The word "formal" encapsulates Bohr's decade-long struggle with the quantum riddle. The new mechanics for these physicists was to be more a direct result of applying symbolic analogy to pre-existing mathematical forms than an entirely novel construction from scratch or an empirical generalization from phenomena. This clear conception of the theory's objective in turn contributed to Heisenberg's major breakthrough in the establishment of the new mechanics.

Heisenberg's (1925) paper, justly titled "quantum-theoretical reinterpretation (*Umdeutung*) of kinematic and mechanical relations," is an attempt to extend symbolic analogy to kinematic quantities and the equations of motion. A similar technique is applied to the Fourier representation of the classical position $x(t)$ by the replacement of C_τ and ω with quantities depending on two discrete values of n and $n-\tau$. The resulting expression is a two-dimensional ensemble of quantities whose sum is not readily calculable. Instead of trying to take its sum, Heisenberg goes on to regard the whole ensemble of quantities as representing a quantum-mechanical position of the electron. Then he proceeds to perform similar translations of higher order quantities $x^2(t)$, $x(t)y(t)$, etc., which led to the first quantum-mechanical equation and solution of an anharmonic oscillator.

In Heisenberg's paper, Bohr witnessed far greater utility of the correspondence principle than hitherto suspected. Heisenberg's "manifolds of quantities," symbolizing the possibilities of state transitions, extend the formal validity of the correspondence principle into the new mechanics. Moreover, Heisenberg's novel approach, while avoiding "difficulties attached to the use of mechanical pictures," attempts to "transcribe every use of mechanical concepts" (5.280), i.e., transcribe every classical formula into the quantum language. This transcription project was nearly complete with the "three-men paper" of 1926. The three men filled in the missing pieces of Bohr's principle, without recourse to a picture or mechanism. As they put this idea, symbolic quantum relations are not "amenable to a geometrically visualizable interpretation . . . in terms of the familiar concepts of space and time." Their dictum that in the new mechanics "symbolic quantum geometry goes over into visualizable classical geometry" essentially captures the contrast between symbol and intuition first suggested by Bohr's correspondence principle earlier. Therefore, it comes as no surprise when they regard the matrix mechanics as an "exact formulation of Bohr's correspondence considerations" (Born, Heisenberg, & Jordan 1926, p. 322).

To answer the question posed earlier, to what extent Kant's notion of analogy is reflected in Bohr's principle, it is rather in this extended sense from formal to symbolic analogy that we can fully appreciate Bohr's Kantianism. The transition from formal to symbolic analogy, except for its final product, went unnoticed by Bohr's contemporary physicists, but its decisive importance for the creation of the new mechanics cannot be underestimated. As much as doctrinal contributions, the entire way of formulating conceptual issues in the old quantum theory is essentially a Kantian problematic. What was initially just a conceptual possibility in Bohr's mind was soon taken up by quantum physicists. Analogy (*Analogie* or *analog*) in the above sense, for example, appears nine times in Heisenberg's *Umdeutung* paper. Similar observations hold for the concept of "intuition" (*Anschauung* or *anschaulich*) in early papers on quantum mechanics. All these concepts, analogy, symbol, picture, intuition, etc., and characteristic interplay between them, should be understood in the Kantian spirit. The emergence of new mechanics is unintelligible unless we take into account Bohr's Kantian philosophy before 1925.

4 The Problem of Scientific Rationality

Thus far, I have narrated, from a Kantian perspective, a philosophical history of how quantum mechanics emerged from Bohr's correspondence principle. One advantage such a narrative has over the usual case histories of science is that it makes the whole quantum revolution philosophically intelligible by integrating parallel developments in scientific philosophy. Scientific philosophy here is taken in its broad sense, not as a discipline of professional philosophers, but as an activity of physicist-philosophers like Bohr. What is the contemporary relevance of the revised historiography, or what is its implication for general philosophy of science? I'd like to address this issue with the problem of scientific rationality during paradigm shifts, i.e., the crisis of rationality Kuhn has been accused of generating.

The quote opening my paper, accusing a philosophy built on the misguided history, was originally directed at Kuhn's opponents. I wonder whether the same criticism might be leveled at Kuhn himself. For a misguided history will pervert any image of science modeled on it, and Kuhn's is no exception. Few other issues in the field illustrate this point better than the problem of rationality. The problem and Kuhn's challenge are as much alive as they were raised some forty years ago. For example, van Fraassen regards it as the central problem of his empiricist epistemology and attempts to answer it with his notion of "bridled irrationality." Revolutionary changes "are not rational because they are rationally compelled; they are rational exactly if they are rationally permitted" (van Fraassen 2002, p. 92). However, in this weak notion of rationality, I see no essential advance over Kuhn, who famously listed several values shared by proponents of rival paradigms. These values are certainly stronger than the notion of bridled irrationality, but they are not strong enough to resolve persistent doubts about the rationality of theory change.

In my view, to meet Kuhn's challenge in his own terms, the only option is to show that revolutionary changes are rationally compelled to a stronger extent than any of the above notions would suggest. The quantum revolution is an ideal candidate for this, not only because it is one of two major shifts in modern physics, but above all, more than any other revolutionary episodes in the history of science, it is a testing case for Kuhn's view of scientific rationality.

Thus "the Bohr research programme," as Lakatos calls it, figures crucially in his debate with Kuhn on scientific rationality (Musgrave and Lakatos 1970). In a manner typical of him, Lakatos claims that Bohr's research programme entered into a degenerating phase in the early 1920s, whereas there arose a rival, progressive research programme of de Brolie-Schödinger's wave mechanics. Bohr's correspondence principle is an "ad hoc stratagem ... the only purpose of which is to hide the 'deficiency'" of the old quantum theory. "Bohr tried in 1922 to lower the standards of scientific criticism" and a similar move after 1925 "led to a defeat of reason within modern physics and to an anarchist cult of incomprehensive chaos" (Ibid., pp. 142–144). In this way, Lakatos illustrates how a philosopher can grossly simplify and misrepresent a revolutionary episode to fit it into his preconceived scheme of scientific rationality.[5] Considering the substantial contributions of the correspondence principle to the quantum revolution and the first-hand acknowledgements by quantum physicists, there is absolutely no way that you can understand Bohr's principle in this way. Perhaps Lakatos hopes to relegate the entirety of matrix mechanics to a part of the degenerating programme, but such an extreme view would convince few historians of quantum theory.

What Lakatos calls the degenerative phase of Bohr's programme after 1922 is for Kuhn a "case book example of crisis," which was brought to an end by matrix mechanics three years later. In this regard, he emphasizes the revolutionary potential of puzzle solving the "creative functions of normal science and crisis." What brought about the resolution of the crisis were not conscious efforts of physicists directed to it, but a normal research directed to some technical puzzles of the old quantum theory. As Kuhn puts it, the "degenerative phase of the old quantum theory provided both occasion and much detailed technical substance" for the matrix mechanics (Ibid., pp. 258). In this regard, he mentions virtual oscillators and Kramers's dispersion formula as normal scientific efforts. However, Bohr's philosophical interventions in both of these, which Heisenberg acknowledged to hold the key to the quantum riddle, were left entirely out of view.

My impression of Bohr is a little different. Feeling of crisis was particularly acute in Bohr, who constantly worried about the physical basis of the old quantum theory. The source of the difficulty was also obvious. According to Heinsenberg Bohr "suffered from the impossibility of penetrating into this very 'unanschaulich' [*unvisualizable*], unreasonable behaviour of nature" (Mehra and Rechenberg 1982, p. 150). That Bohr and Bohr alone was conscious of the crisis does no good for Kuhn. First, there was no feeling of crisis at the community level. The feeling was shared neither by the Munich school of Sommerfeld nor by the Göttingen school of Born, both of which mostly focused on solving technical puzzles of a mathematical character in the old quantum theory (Ibid.). However, it was Bohr for whom the conscious realization of the crisis offered the opportunity of rational resolution. It was not that Bohr was driven, as it were, almost in desperation to philosophical speculations in the face of mounting anomalies. Rather, Bohr was experimenting with various philosophical options as illustrated in the BKS theory. Contra Kuhn, these activities were from the beginning directed precisely to the conceptual difficulties of the old quantum theory, whose resolution was prerequisite for the foundation of a new mechanics. Bohr was opening various conceptual possibilities for the new mechanics, one of which (not the BKS picture, but the symbolic analogy) was soon to be actualized and embodied in it. The impetus behind the revolution was, as Kant puts it, "reason commanding nature to answer questions formulated in accordance with its principle." Those conceptual possibilities made available by the correspondence principle loosened the grip intuitive pictures had on physicists, thus facilitating the

[5]Therefore, Kuhn is right in his note (Musgrave and Lakatos 1970, p. 256) where he points out deficiencies of a philosopher's history like Latakos's. Mine may not be free from the same accusation, but at least it is not as liberal distortions as Lakatos's and claim to be a "rational reconstruction" to a similar extent to Kuhn's.

transition to the new paradigm. Thus we should attribute the "creative functions" primarily to Bohr's philosophical principle, not to some concrete puzzling solving within the old quantum theory.

Bohr's philosophical notion of analogy provides a case for a meta-scientific intervention in a theory in the making and thus room for prospective, as opposed to retrospective, rationality.[6] We do not invoke the correspondence principle as a justification of the paradigm shift ex post facto, but as the very momentum behind it. Without such activities, the revolution is entirely unpremeditated or unprepared, and this is the main trouble with Kuhn's notion of rationality. For Kuhn, there is no independent source or standard of rationality external to normal sciences, no "Archimedean platform" beyond shifting paradigms. However, if we view the correspondence principle exclusively as a paradigm for the old quantum theory, then we cannot account for the fact that it is a principle of quantum theory valid for both old and new mechanics. Therefore, in the quantum revolution, the correspondence principle as formal or symbolic analogy should be regarded as a common platform, or rather, a meta-scientific resource, on which both old and new mechanics could draw.

A similar observation was made by Friedman about Kuhn's trouble with scientific rationality in the context of the theory of relativity. According to Friedman (2004, p. 90) while the rationality of the relativistic revolution was mediated by scientific philosophy of the previous century, Kuhn "left out this parallel history of scientific philosophy," and this accounts for why Kuhn could not resolve the problem of rationality engendered by his historiography. Of course, in the old quantum theory, we cannot deny that normal researches provided much needed heuristics and technical sophistication for the upcoming revolution, and the strength of Kuhn's historiography is his detailed treatments of these issues. However, by leaving out the meta-scientific dimension or philosophical interventions during the revolutionary transition, Kuhn makes the latter unintelligible to philosophically minded readers. My point here is that if philosophical understanding has any relevance to our understanding of quantum theory, normal science and the ensuing crisis are not all there are to the old quantum theory.

In dealing with the problem of scientific rationality, philosophers have mostly focused on the evaluation of a theory or a hypothesis in light of data or evidence. The main question for them was whether it is rational to accept, maintain, revise, or reject a theoretical hypothesis given a certain body of evidence. Various formal or computational models and theories of scientific rationality were suggested, prominent among them being Bayesian approaches (e.g., Salmon 1990). However, few have questioned the evidential ground for Kuhn's historiography, and a few who did so like Lakatos have failed abjectly.

About the troublesome problem of rationality, I feel we are in a similar position to where Hacking (1983) stood about the problem of reality. The skeptical part of his book concerned whether the problem of reality can ever be resolved at the level of theory and representation. A similar reservation can be raised about the problem

[6] See Friedman (2001, p. 101; 2002, pp. 185–6) for this important distinction.

of rationality, about whether it can be resolved at the level of theory and evidence. Kuhn's point that the theory choice "cannot be resolved by proof" seems inexorable. Rather, to meet Kuhn's challenge in his own terms, we should explore anew the very historicity of the concept of scientific rationality and the proper role of reason in the construction of a new paradigm. As one author puts it, "data drawn from the history of science somehow *constitute* or are *evidential for* the concept of rationality" (Matterson 2008, original emphasis). By doing *more* history, philosophically informed history, we may recover reason in action that was buried in technical puzzle solving in Kuhn's historiography. In this way, a philosophical history of quantum theory can contribute to reestablishing once renounced rationality of theory change.

References

(volume # .page #) refers to Bohr, N. (1972–2007). *Collected works*, general editor L. Rosenfeld (Vols. 1–12), Amsterdam: Elsevier.

Born, M. (1924). Über Quantenmechanik. *Zeitschrift für Physik*, *26*, 379–395, translated in (B. L. van der Waerden (Ed.), *Sources of quantum mechanics*, pp. 181–198).

Born, M., Heisenberg, W., and Jordan, P. (1926). Zur Quantenmechanik II. *Zeitschrift für Physik*, *35*, 557–615. translated in B. L. van der Waerden (Ed.), *Sources of quantum mechanics*, (pp. 321–385). Amsterdam: North Holland Publishing Company, 1967.

Chevalley, C. (1994). Niels Bohr's words and the Atlantis of Kantianism. In J. Faye & H. Folse (Eds.), *Niels Bohr and contemporary philosophy, Boston studies in the philosophy of science* (Vol. 153, pp. 33–55). Dordrecht: Kluwer.

Chevalley, C. (1995). Philosophy and the birth of quantum theory. In K. Gavroglu, J. Stachel & M. Wartofsky (Eds.), *Physics, philosophy, and the scientific community, Boston studies in the philosophy of science* (Vol. 163, pp. 11–37). Dordrecht: Kluwer.

Darrigol, O. (1992). *From c-numbers to q-numbers: The classical analogy in the history of quantum theory*. Berkeley: University of California Press.

Friedman, M. (2001). *Dynamics of reason, the 1999 Kant lectures at Stanford university*. Stanford: CSLI Publications.

Friedman, M. (2002). Kant, Kuhn, and the rationality of science. *Philosophy of Science*, *69*, 171–190.

Friedman, M. (2004). Philosophy as dynamic reason: The idea of a scientific philosophy. In H. Carel & D. Gamez (Eds.), *What philosophy is: Contemporary philosophy in action* (pp. 73–96). New York: Continuum.

Hacking, I. (1983). *Representing and intervening, introductory topics in the philosophy of natural science*. Cambridge: Cambridge University Press.

Heisenberg, W. (1925). Über quantentheoretische Umdeutung kinematischer und mechanischer Beziehungen. *Zeitschrift für Physik*, *33*, 879–893, translated in (B. L. van der Waerden (Ed.), *Sources of quantum mechanics*, pp. 261–276).

Hertz, H. (1956) [1894]. In D. Jones & J. Walley (Eds.), *The principles of mechanics, presented in a new form, trans*. New York: Dover.

Høffding, H.(1905). On analogy and its philosophical importance. *Mind*, *14*(54), 199–209.

Lee, J. (2006). Bohr vs. Bohm: Interpreting quantum theory through the philosophical tradition (Ph.D. dissertation, Indiana Univeristy, 2006).

Matterson, C. (2008). Historicist theories of rationality. In E. N. Zalta (Ed.), *The Stanford Encyclopedia of Philosophy* (Spring 2009 Edition).

Mehra, J., & Rechenberg, H. (1982). *The historical development of quantum theory*, Vol. 2: *The discovery of quantum mechanics*, 1925. New York: Springer.

Musgrave, A., & Lakatos, I. (1970). *Criticism and the growth of knowledge*. Cambridge: Cambridge University Press.

Pringe, H. (2007). *Critique of the quantum power of judgement: A transcendental foundation of quantum objectivity*. Berlin: Walter de Gruyter.

Rosenfeld, L. (1973). The wave-particle dilemma. In J. Mehra (Ed.), *The physicist's conception of nature* (pp. 251–261). Dordrecht: Reidel.

Rozental, S. (1967). *Niels Bohr: His life and work as seen by his friends and colleagues*. New York: Wiley.

Salmon, W. (1990). Rationality and objectivity in science or Tom Kuhn meets Tom Bayes. In C. Savage (Ed.), *Scientific theories, minnesota studies in the philosophy of science* (Vol. 14, pp. 175–204). Minneapolis: University of Minnesota Press.

Sharrock, W., & Read, R. (2002). *Kuhn: Philosopher of scientific revolution*. Cambridge: Polity Press.

Tanona, S. (2002). *From correspondence to complementarity: The emergence of Bohr's copenhagen interpretation of quantum mechanics* (Ph.D. dissertation, Indiana Univeristy, 2002).

van Fraassen, B. (2002). *The empirical stance*. New Haven: Yale University Press.

Wheeler, J., & Zurek, W. (1983). *Quantum theory and measurement*. Princeton: Princeton University Press.

Understanding Galileo's Inquiries About the Law of Inertia

Jun-Young Oh, YooShin Kim, Chun-Hwey Kim, Byeong-Mee Min and Yeon-A Son

Abstract The purpose of this research is to gain a better understanding of the role of abstraction and idealization in Galileo's scientific inquiries about the law of inertia, which occupies an important position in the history of science. We argue that although the terms "abstraction" and "idealization" are variously described in the recent literature, the concepts must be adopted to highlight important epistemological problems. In particular, we illustrate the importance of abstraction and idealization for the formation of the law of inertia by establishing a distinction between two types of entities: quasi-ideal entities and idealized entities. These theoretical laws should therefore be justified, using deduction and induction, through quasi-idealized entities based on data from the everyday world.

1 Introduction

The notions of idealization and abstraction, which are commonly used in the thought processes of scientific knowledge construction, can be traced back to Galileo (Portides 2005). Nola (2004) argues that a clear distinction must be made between idealization and abstraction, so that the use of abstraction and idealization could be considered valid once the two notions are distinguished (Cartwright 1989; Suppe 1989; Portides 2005).

According to Nowak, "[in] brief, the Galilean breakthrough consisted in the introduction of the method of idealization in physics":

J.-Y. Oh (✉)
Hanyang University, Seoul 133-791, Republic of Korea
e-mail: jyoh3324@hanyang.ac.kr

Y. Kim
Pusan National University, Pusan 609-735, Republic of Korea

C.-H. Kim
Chungbuk National University, Cheongju 361-768, Republic of Korea

B.-M. Min · Y.-A. Son
Dankook University, Gyeonggi 448-701, Republic of Korea

L. Magnani et al. (eds.), *Philosophy and Cognitive Science II*,
Studies in Applied Philosophy, Epistemology and Rational Ethics 20,
DOI 10.1007/978-3-319-18479-1_11

> The Galilean revolution consisted in making evident the misleading nature of the world image that the senses produce. We only see phenomena that are the joint effect of all the relevant influences. As a result, senses do not contribute in the slightest to the understanding of the facts. In order to understand phenomena the work of reason is necessary which selects some features of the objects through idealization and in their idealized models recognizes some other features of the empirical originals. These models differ a great deal from their sensory prototypes; what is more, they present images of hidden relationships that could not be grasped with the aid of experience at all. (Nowak 1994, in Nola 2004, p. 350).

The hidden causes of observable events cannot, by definition, be perceived by our senses but can be revealed by idealization and abstraction. Therefore, to understand important scientific cognitive processes, the exploration of how idealization takes place in science is mandatory. The case of Galileo is central: he used a dyadic approach, combining the composition of idealizations for theories or models and the generation of observation-based predictions, about which we can only make approximations using our theories or models. The present study focuses on this dyad.

First of all we will present the development of a typical scientific inquiry based on different level worlds. We will then describe in detail Galileo's discovery of the Law of Inertia. How did Galileo use the abstraction and idealization strategies in formulating the law of inertia? The answer can be found by showing how Galileo went beyond Aristotle's scope of observation through the use of thought experiments. Moreover, how did Galileo justify the law of inertia that he formulated? Answering this question will allow us to show how the proof of his conclusion was also produced through thought experiments. A final question also arises in relation to heliocentrism: How did Galileo's strategies concerning Copernicus' work support the heliocentric hypothesis?

2 Background

2.1 Abstraction and Idealization

Modern philosophy of science distinguishes between abstraction and idealization. As an important activity in constructing models and theories, abstraction comprises processes of forming general concepts out of individual instances.

Psillos describes abstraction as the removal of certain characteristics, properties, or features of an object or system that are not related to the aspects of behavior of the object under investigation (Psillos 2007, p. 6). According to Cartwright, in the case of abstraction, we subtract concrete facts about objects, including perhaps the details of their material composition, and—of particular importance—we eliminate interfering causes (Cartwright, in Ladyman 2002, p. 260). By contrast, Cartwright defines idealizationas the theoretical and experimental manipulation of concrete situations to minimize or eliminate the effects of certain features of an object. For example, we obtain a frictionless plane through idealizing the real surface, and then we re-introduce the appropriate mathematical concept, the friction coefficient.

Bhaskar (1979/1998) interprets abstraction as follows:

> Abstraction according to its traditional meaning is focusing upon certain aspects of something to the (momentary) neglect of others. It is a process of focusing on some feature(s) of something(s) while others remain in the background (Bhaskar 1979/1998, p. 170).

In other words, abstraction allows us to omit less relevant details, as we understand them, and thus to increase our capacity to identify an object's essential features. Hypotheses about idealizations can not be obtained by induction, by simple enumeration, or by the methods of agreement and difference. The scientist must intuit which properties of phenomena form the proper basis for idealization and which properties may be ignored (Losee 2001, p. 49).

Nola (2004) succinctly differentiates between abstraction and idealization as follows:

> The term "idealization" will be used differently. In the case of abstraction, an object is still a real object with property P, but we ignore property P for certain purposes, such as whether it is a property with which our theory deals. But in the case of idealization, we do not merely ignore a property; we regard P as a property that the object *definitely does not possess*. (Nola 2004, p. 357).

According to Nola, in idealization we do not ascribe ontological status to the removed properties of the object, whereas in abstraction we ascribe an ontological status to removed properties. Moreover, abstraction means consciously ignoring certain properties for certain purposes.

Nola's idealization means considering the properties that the object does not definitely possess rather than simply ignoring certain features of the object deliberately. Generalizing for an infinite amount of unobservable data based on a finite amount of observed data, instead of ignoring certain features, is an important example of formal idealization.

In our research concerning Galileo's work, abstraction and idealization are distinguished as follows:

> **Abstraction**: In scientific activity, the notion of abstraction is essential. Abstraction allows the scientists to focus on an object's particular properties—and their operations—through isolation, controlling and removing any other properties present in the concrete circumstances.
>
> **Idealization**: Idealization is the consideration of properties that the object definitely does not possess in a physical system, using Galileo's thought experiments; the notion of abstraction, by contrast, deliberately ignores certain features the object possesses in concrete circumstances, while others remain in the background

By extrapolating from a series of phenomena, we formulate an idealization. For example, the concept of free fall in a vacuum can be obtained through extrapolation based on the observations of the behavior of a falling object in a series of fluids of decreasing density. Creative imagination plays an important role in obtaining results by means of idealization.

Abduction and Retroduction: The abductive process simultaneously infers the rule and the case from a known fact (i.e., the result) that requires explanation. Abduction is an expansive cognitive process in the sense that it yields a novel hypothesis (amplitude). How can abduction be a form of inference distinct from deduction and induction (as the unfettered play of amusement or as a response to a surprising fact) and also a form of recursive analysis that includes deduction and induction? Referring to the concept of abduction as amusement and to that of recursive analysis as retroduction can eliminate much of the confusion surrounding abduction.

The distinction between the pre-trial and post-trial evaluations of hypotheses is included in the H-D method. For example, Whewell required the use of a hypothetical theory to "explain phenomena which we have observed" and "foretell phenomena which have not yet been observed", indicating that these phenomena are "of a kind different from those which were contemplated in the formulation of our hypothesis" (Whewell 1847, pp. 62–65).

According to Rescher (1978), Peirce sees qualitative induction as an evolutionary process of variation and selection. Two component processes are involved here, as we have seen:

1. Hypothesis production or abduction: the purely conjectural proliferation of a plethora of alternative explanatory hypotheses that are relatively plausible.
2. Hypothesis testing or retroduction: the elimination of hypotheses on the basis of observational data (Rescher 1978, p. 8).

The result of the overall process is that science proceeds by the repeated elimination of rival hypotheses in favor of one preferred candidate. Each stage of the abduction-retroduction cycle reduces a cluster of conjectural hypotheses to an accepted theory.

2.2 The World According to Galileo's Scientific Inquiry Procedure

In this section, we propose a model of the scientific inquiry process which is useful to interpret the one adopted by Galileo. We stratify the ontological world into four layers—the empirical world, the event world, the theoretical world, and the idealized world—to analyze Galileo's scientific inquiry.

The empirical world: The empirical world may be roughly defined in relation to the other worlds. The empirical world is the world perceived through the sense organs of a cognizing subject who accommodates the stimulations of the external world. In this world, the cognizing subject is in a state of ignorance with regard to the regularities among perceptions, but these regularities, the features of events in this world, are grasped by the cognizing subject via perception of the external world and are produced through arranging and individuating them.

The theoretical world: The theoretical world comprises the description of causal forces and mechanisms operating on objects outside the cognizing subject. The elements of the theoretical world, therefore, correspond to the elements of the external world of the cognizing subject. Although certain objects exist, the causal forces of those objects may not come into effect because of inappropriate conditions and consequently may not produce suitably detectable results. The theoretical world may not exactly reflect the external world but approximately conforms to it. Theories always aspire to describe the external world satisfactorily.

> **The idealized world**: The idealized world is composed of idealized entities that are mental constructions of the scientist, produced by idealization and based on objects in the event world. A vacuum is an ideal entity, and free fall in a vacuum is the ideal behavior of an ideal entity. Idealization requires creative imagination rather than a simple enumeration-induction and consistence-difference method.

We argue that experiments involving idealization are at the core of Galileo's arguments against the Aristotelians. In our study, the idealized world belongs to the theoretical world. The theoretical model is a set of idealized objects and idealized relations among them, obeying idealized laws (Nola 2004, p. 360).

In Galileo's writings, thought experiments and real experiments are deeply intertwined and often indistinguishable. Some of his experiments proved to be only imaginary (Cohen 1950/1993; Cushing 1998; Dijksterhius 1986, pp. 81–84). All of these experiments amount to the same thing—idealized experiments—in contrast to real experiments in the modern sense of the term (Galili 2009). Here, we can see that Galileo's thought experiments constitute a significant strategy of connecting and coordinating real experiments and theoretical models involving idealized entities.

Recently, Fernández-González (2013, p. 1727), proposed an ideal level and a quasi-ideal level for both physics and chemistry:

> Idealized entities are thus archetypes of real world objects. Unlike Plato's ideal entities, which are eternal and immutable, idealized objects are mental constructions of the scientist, based on real objects. Quasi-ideal entities are those real world entities whose characteristics most closely approximate those of idealized entities since they are created with that intention (e.g., the balls used by Galileo that imitate geometrical spheres). This quasi-ideal world is an almost perfect reflection of the ideal world. Thus, if the level of precision required is not very high, a quasi-ideal system can behave as though it were ideal. Actually, the ideal world is part of the theoretical world, where complex structures such as theories and models reside.

By using thought experiments, Galileo was able to extend the scope of the concept of experience without distorting its validity. These thought experiments enabled him to discover things about motion that could not be discovered using common experience and simple inference (Gower 1997, pp. 31–32). For Galileo, thought experiments secured a link between the real experiments that led him to his belief in the principles of a new science of motion and the common experiences that he would need to appeal to in order to justify those principles to his readers (Gower 1997, p. 33). A scientific inference based on abstraction and idealization for the generation of hypotheses involves the deduction-induction method for justification of the hypothesis (see Magnani 2001, 2009; Oh 2012, 2014) and for the evaluation of hypotheses.

In this study we are considering three distinct worlds: the empirical world is separate from the idealized world, which is embedded in the theoretical world, thus giving the appearance of two worlds. We locate the theoretical world, including the idealized world, at the highest level and describe its members as theoretical entities (theories, models) and idealized entities. We call those entities situated at the highest level of the empirical world, close to the idealized world, quasi-idealized entities.

2.3 Galileo's Scientific Inquiry Procedure about the Law of Inertia

The generation and formulation of hypotheses based on abstraction and idealization.

In the present study we are analyzing the generation of hypotheses through abstraction and idealization. The processes of abstraction and idealization are as follows: first, we formulate new hypotheses through prevailing knowledge about a theoretical physical world, a process known as "abduction". We then make use of an idealization strategy based on thought experiments to generate an idealized physical world (see Fig. 1).

The arrows directed upward symbolize the generation or formulation of hypotheses, and the arrow downward symbolizes the verification or justification of the generated hypotheses. The dotted arrows mark the border between the theoretical world (or conceptual world) and the event world.

The relationship between the ideal world and the quasi-ideal world is somewhat closer. The quasi-ideal world is the closest approximate reflection of the ideal world, but the quasi-ideal world has no meaning without the existence of the ideal world, which is its referent.

Following Kuhn (1970), we also believe that Galileo observed the same phenomena through different paradigms. In fact, observation is theory-dependent because, when we observe phenomena, we are influenced by what we already "know" and by

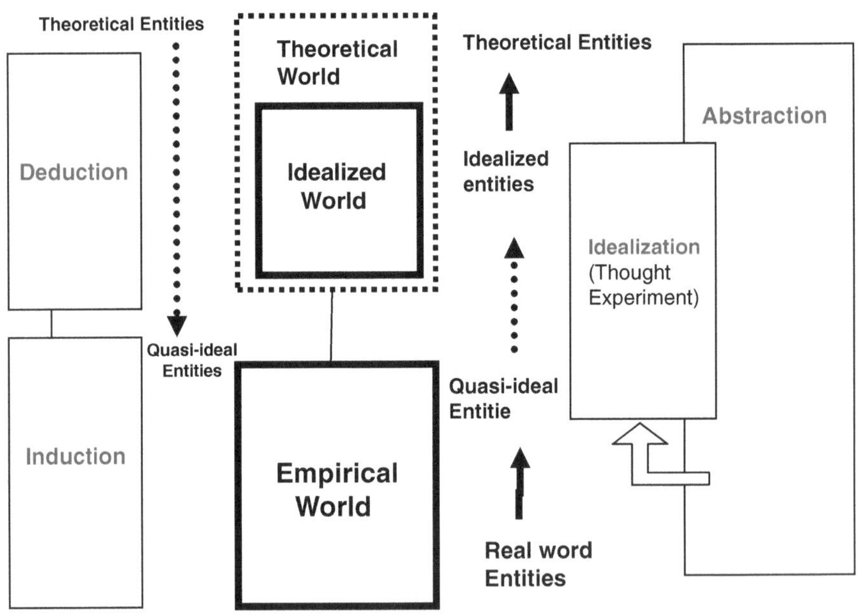

Fig. 1 Galileo's scientific inquiry procedure about the law of free falling motion

background thoughts. Galileo poured all his efforts into developing the heliocentric theory and he accepted Aristotle's theory about the natural motion of the essential elements of the universe. Also the influence of Neo-Platonism in his abductive inferences can be easily detected.

The testing and verification of hypotheses based on deduction-induction

Deduction: Grosseteste and Roger Bacon (Losee 2001, p. 49) improved the Method of Composition by suggesting the deduction of consequences (quasi-ideal entities in the event world) that were not initially included in the data in order to induce explanatory principles (theoretical entities in the theoretical world).

Induction through Experimental Confirmation: Grosseteste and Roger Bacon (Losee 2001, p. 50) also added to the method of resolution and composition a third stage in which the conclusions reached are further tested experimentally (quasi-ideal entities in the event world). In fact, only quasi-ideal entities in the event world appear in experimental situations. Because of the range of possible forces at work, it is difficult to construct scientific experimental situations in which natural phenomena can be explored, so to speak, in their entirety.

3 Galileo's Formulation and Justification of the Law of Inertia

Logic, considered as a tool assisting the search for truth, had different meanings for Simplicio (Aristotle) and Salviati (Galileo). Simplicio regarded Aristotelian logic itself as the absolute authority and the correct tool for finding truths. In contrast, Salviati believed that the true logic that reveals natural knowledge resides in the proofs of mathematics and geometry, which are the "languages of Nature". Garrison (1986) referred to this activity of explaining the empirical world using the ideal world as "secondary idealization". This is the Galilean solution to the problem of generalization that has become common in the physical sciences (see Chalmers 1990, p. 35).

3.1 Galileo's Strategies for Formulating the Law of Inertia

First Stage: abstraction based on abductive strategies

Salviati: [...] Now tell me, what do you consider to be the cause of the ball moving spontaneously on the downward inclined plane, but only by force on the one tilted upward?

Simplicio: That the tendency of heavy bodies is to move toward the center of the Earth and to move upward from its circumference only with force; now the downward surface is that which gets closer to the center, while the upward one gets farther away (Galilei 1967, p. 148).

However, if an external force is not continuously added to the horizontal surface, the moving object eventually stops. Feyerabend (1975) argued that Galileo ascribed a special position to motions that are neither natural nor forced:

> It must be assumed that the "neutral" motions, which Galileo discusses in his early dynamical writings, may be forever or at least for periods comparable to the age of historical records. And they must be regarded as "natural," in the entirely new and revolutionary sense that neither an outer nor an inner motor is needed to keep them going (Feyerabend 1975, p. 95).

In an exchange between Simplicio and Salviati, two characters in his Dialogues, Galileo argues that because a smooth ball projected down an incline would accelerate and continually increase its speed and because one projected up an incline would decelerate and continuously slow down until it stopped, then a ball projected along a horizontal surface would continue to move along this surface with undiminished speed. In some passages, Galileo seemed to consider natural, or inertial, motion to be a form of motion that neither rises nor falls but that always remains equidistant from the center of the Earth (McMullin 1985).

Aristotle had held that Earth could not be moving because if it were, objects such as birds, falling stones, and clouds would be left behind as Earth moved along its way. Galileo defused this objection to the sun-centered idea with experiments that almost single-handedly overturned the Aristotelian view of physics. In particular, he used experiments with rolling balls to demonstrate that a moving object remains in motion. This insight explained why objects that share Earth's motion through space—such as birds, falling stones, and clouds—should stay with Earth rather than falling behind as Aristotle had argued. Thus, Galileo formulated the inertial law of the terrestrial world from Aristotle's celestial natural motion through what it is now called rule-forming abduction.

Abduction: Rule-forming "Abstraction"

Aristotle divided motion into the "natural" motions and the "violent" motions. Natural motions were those motions that objects naturally made: objects on Earth fell towards the center of the Earth. Heavenly objects naturally moved in circles. Violent motion was *anything* other than this. Therefore, picking up a rock was considered a violent motion.

> Because motions in the horizontal plane with no friction are not going down and going up based on the natural motion of Aristotle, it is possible for the motion of rolling to continue. If the hypothesis that neutral motion without any friction at the same distance from the center of the Earth is equal to natural motion in the celestial world is correct, and if the objects hold some speeds, then an object projected along a horizontal surface equidistant from the center of the Earth will continue to move along this surface with undiminished speed (theoretical entities). Therefore, there are good reasons to believe that an object projected along a horizontal surface equidistant from the center of the Earth will continue to move along this surface with undiminished speed.

In addition, the notion of eternal straight motion, another hypothesis in the horizontal plane, was eliminated based on Aristotle's finite universe **(Retroduction)**.

Galileo contradicted the strict separation between the celestial and terrestrial worlds of Aristotle. In addition, he developed them into a claim that natural motions ofthe vertical rise and fall in the terrestrial world are the same movement as the circular movement in the celestial world. Because Galileo cannot deviate completely from Aristotle's views regarding celestial motion in perfect circles, his inertial motion was circular rather than the linear inertial motion of Newton.

Second Stage: idealization based on thought experiments

"The removal of friction by thought experiments: This produces ideal objects that act in conditions that are also ideal".

If the hypothesis that an object experiences friction on a flat surface is correct and if the flat surface is infinite and an object is pushed with the same force on a surface that is increasingly smooth, then what motion will the object follow?

(But) In fact, the object stopping further away on an increasingly smooth flat surface (quasi-ideal entities) and eventually moving infinitely farther away is possible as an ideal limit value. Therefore, in an idealized situation (idealized entities) that ignores air resistance, the flat surface can be made smoother and smoother until eventually the object continues without stopping.

We believe that Galileo imagined a smooth flat surface becoming smoother and smoother as a quasi-ideal entity in an experimental situation, with the surface finally becoming frictionless. He regarded this plane as an ideal entity. As an example of such an entity, he cited the surface of a smooth water plane, and he saw the infinite frictionless plane as an idealized entity. He, thus, advanced his idealization strategy through two types of entities: idealized and quasi-idealized.

For idealized entities, no friction exists between objects and surfaces. If all resisting effects could be removed, the object would continue in a steady state of motion indefinitely in a theoretical world. Theoretical entities (models or theories) in a theoretical world consist of idealized entities. Galileo is close to the law of inertia in the Newtonian sense, but this point does not imply that Galileo reached the law of inertia in the Newtonian sense, only that he developed an idealization strategy through creating quasi-ideal entities.

Under his idealization strategy, if a plane and an object are both highly polished, the object, given the same initial speed, will slide farther before coming to rest. On a smooth layer of ice (the closest approximation of an idealized entity and therefore a quasi-ideal entity in the event world), the object will slide farther still. Galileo reasoned that if all resisting effects could be removed, the object (an idealized entity) would continue in a state of motion indefinitely (a theoretical entity).

We believe that Galileo imagined a smooth, flat surface becoming smoother and smoother as a quasi-ideal entity in an experimental situation, with the surface finally becoming frictionless. He regarded this plane as an ideal entity. As an example of such an entity, he cited the surface of a smooth water plane, and he saw the infinite frictionless plane as an idealized entity. He,thus, advanced his idealization strategy through two types of entities: idealized and quasi-idealized.

3.2 The Process of Justification

Such inertia has been applied in the tower argument, which supports the Copernican theory of the Earth's motion (Finocchiaro, 2010) and is consequently proved: although a ship on a smooth sea is motionless, once an external force strikes it, the ship will move around the surface of the Earth continuously. Of course, no resistance and no external force are present.

According to Cohen (1985), in Galileo's experimental test the essence of what is called both the mathematico-experimental and hypothetico-deductive method in symbolic terms is displayed:

> What Galileo did was to deduce B from A; he next tested B, and then concluded that A holds. It should be noted, however, that this method does not include a guarantee of A. For instance, it might happen that B could also be that deduced from A'. Additionally, it is assumed that the process of deducing B from A is correct. Traditionally, this means correctness of logical deduction. Galileo's method is to derive B from A by aid of mathematics. Because B is derived from A by mathematics and then tested by experiments, the method can also be called mathematico-deductive. In the seventeenth century, the term "mathematical-experimental" was also used. This method is called "hypothetico-deductive" because we wish to test hypothesis A but cannot do so by experiment. (Galileo's use of this method in relation to the hypothesis $V \propto t$ and the testable deduction $D \propto t^2$ may be found on pp. 88 ff. supra.) (Cohen 1985, p. 207).

> **Deduction**: If the falling body speed (V) of an object is proportional to falling time (t) and, ignoring air resistance, if the object is in free fall without an initial velocity, then it is mathematically induced that the fall distance (D) is proportional to a square of the time (t^2).

> **Induction**: As mathematically predicted, the result showed that V is proportional to $t(t^2)$. Thus, the hypothesis that the falling time of an object in free fall, ignoring air resistance, is proportional to the square of the falling time is supported.

Third Stage:

"The Deduction-Induction Cycle"

In Galileo study, it was by an experiment that, in parabolic motion of an object that was dropped from a mast on a ship like the following, the horizontal motion components of an object that is mathematically induced equals the velocity of the ship. In other words, Galileo realized for the first time that physical quantity that has a direction, that is, vector, can be decomposed into components that are at a right angle to one another.

One of the arguments against the heliocentric hypothesis of Copernicus was the tower argument. How did Galileo counter it? Galileo's study experimentally proves that, from the parabolic motion of an object dropped from the mast of a moving ship, the speed of the boat is, in fact, a mathematically derivable element of the constant horizontal motion of the object.

(If) the law of inertia (a theoretical entity), which states that all objects that are not provided with a force for motion have inertia in motion, is correct, (and) if an object is dropped from the mast of a ship that moves at a constant speed and has no air resistance (quasi-ideal entities), (then) the object must fall directly below the mast due to the mathematically derived horizontal velocity of the object.

Several well-known passages in Galileo's *Dialogue Concerning the Two Chief World Systems* relate to this:

> Salviati: Now as to that stone which is on top of mast; does it not move, carried by the ship, both of them going along the circumference of a circle about its center? Consequently, is there not in it an ineradicable motion, all external impediments being removed? And is not this motion as fast as that of the ship? (Galilei 1967, p. 148).
>
> Salviati: [...] for you yourself have already granted the resistance to be against motion which increases the distance from the center, and the tendency to be toward motion which approaches the center. From this it follows necessarily that the moving body has neither a resistance nor a propensity to motion which does not approach toward or depart from the center, and in consequence no cause for diminution in the property impressed upon it. (Inertial law: Theoretical entities) (Galilei 1967, p. 149)
>
> Salviati [...] it may be seen that, at most, the falling body might drop behind if it were made of light material and the air did not follow the ship's motion, but if the air were moving with equal speed, no imaginable difference could be found in this or in any other experiment you please (*Expected Results: Quasi-ideal entities*) [...] Now, in this example, if no difference whatever appears, what is it that you claim to see in the stone falling from the top of the tower, where the rotational movement is not adventitious and accidental to the stone, but natural and eternal, and where the air as punctiliously follows the motion of the Earth as the tower does that of the terrestrial globe? (*Support of the Heliocentric hypothesis*). (Galilei 1967, p. 154, emphasis added).

Induction through Experimental Confirmation and Expansion:

(If) the law of inertia (a theoretical entity), which states that all objects that are not provided with a force for motion have inertia in motion, is correct, (and) if an object is dropped from the mast of a ship that moves at a constant speed and has no air resistance (quasi-ideal entities), (then) the object must fall directly below the mast due to the mathematically derived horizontal velocity of the object.

Following the steps above deduction, Induction,

(And) as expected, the object that was dropped from the mast of a moving ship fell directly below the mast of the ship (quasi-ideal entities). (Therefore), the hypothesis that all objects that are not provided with a force for motion have inertia in motion is supported.

One of the questions left unanswered by the Copernican system of the universe was the following: if the Earth rotates, then why does a ball dropped from a tall tower fall directly below instead of some distance away from the tower? Galileo's response was that the ball shared the Earth's rotation and continued this motion in a horizontal direction even while it was falling down. This response reflects the principle of inertia, according to which all motions maintain their status as long as there is no external interference (Cushing 1998). The same logic that explains why the ball falls directly below the tower on a rotating Earth also explains why a ball falls directly below the mast on a moving ship. Galileo's principle of inertia did not deviate from Aristotle's view that a circular motion is complete and natural, however. The orbit of an inertial motion is a circle because the ball falling from the tower shares the Earth's rotational motion.

According to Cohen (1985), Galileo intended to prove his restricted inertia through this well-known fact:

> It is precisely this point which Galileo wished to prove because he now can explain that a stone let fall from a ship will continue to move around the Earth as the ship moves, and so will fall from the top of the mast to the foot of the mast (Cohen 1985, p. 121).

Galileo recognized this constant motion as a uniform circular motion. The planets absolutely must follow a circular orbit at a constant speed. This is because any change in motion meant that there was a continued application of force affecting the planets, or as Kepler said, a magnetic-like force affecting them. Galileo thought, according to his own law of inertia, that the orbits of the planets must be circular and that no change could occur in the speed of planetary motion (Parisi 2001, pp. 23–24).

Galilean relativity theory says that within the inertial frame, system, uniform motion, and stationary state cannot be distinguished. This means that the absolute space of Aristotle does not exist. In a famous passage from his *Two Chief World Systems*, Galileo says:

> Salviati: How many propositions I have noted in Aristotle that are not only wrong, but wrong in such a way that their diametrical opposites are true, as happens in this instance! But keeping to our purpose, I believe that Simplicio is convinced that from seeing the rock always fall in the same place, nothing can be guessed about the motion or stability of the ship (Galilei 1967, pp. 153–154)

This experiment provided the basis for an interesting and important objection to the Copernican heliocentric hypothesis of the Earth's motion, especially to the Earth's daily axial rotation.

4 Discussion

The "thought experiments" by which Galileo destroyed the Aristotelian dogma that moving objects stop eventually and heavier objects fall faster than lighter objects are classic and typical examples in the field of the science of motion. These experiments are certainly prototypical examples and thus feature prominently in all contemporary studies concerning scientific thought experiments (Brown 1991, 2000; Gendler 1998; Norton 1996). Galileo's thought experiments take two-fold roles for the formation and justification of theory. Galileo started with an analysis of idealized conditions that experience could never provide. Where Aristotle had begun with experience, Galileo began with the idealized case, of which the actual is only an imperfect embodiment. Having defined the ideal, he could then understand the limitations that material conditions entailed. If we start from experience, we are more likely to end up with Aristotle's mechanics, a highly sophisticated analysis of experience. Fundamental to Aristotle's discussion was the principle that all of the objects we encounter on the Earth are made up of some combination of "four elements": air, Earth, fire, and water. The celestial bodies are made of a "fifth element": "aether." The natural motion of a body composed of aether is circular, so that the observed circular motion of the heavenly bodies is their natural motion, just as motion upward or downward in a straight line is the natural motion for a terrestrial object.

Galileo also proposed the application of geometry to the study of terrestrial motions, such as the inertial motion based on abstraction and idealization in this study, through which he implied, ultimately, that Earth becomes a celestial body in a Copernican system. What shocked Aristotelians was that Galileo attempted to apply mathematical schemas to the terrestrial material cases, formerly only applicable to the explanations of the perfect celestial movements, believing that this rule was hidden in the "imperfect" Earth. Galileo thought that imperfection was the basis of reality. The imperfection of a fall of a body began the starting point of dynamics, and the spots sparkling around Jupiter as well as the solar spots on the surface of the sun demonstrated the possibility of the heliocentric theory. With Galileo, matter suddenly existed as "itself." He changed the question of "why" into "how".

According to Aristotle, without an applied external force, an object will move linearly upward or downward in accordance with its natural tendency to reach its original position. Aristotle thought that each of the basic elements had an original location as one of its properties and that movement stopped when an element had reached its original place. Although Galileo accepted Aristotle's theory of natural motion, he argued that a circular motion, that is, the natural celestial motion, could be induced on the ground by a violent motion, such as that caused by the application of a significant pushing or pulling force on the object. The primary weakness of Aristotle's approach was that it was, so to speak, too empirical. This weakness explains why he was unable to produce a mathematical theory of nature. Galileo's major achievement was to dare to describe the world although we do *not* experience it. He stated laws in a way that was outside of direct experience and, therefore, could not be verified by any single observation, but that was mathematically simple. Thus, he opened the road to a mathematical analysis that breaks down the complexity of actual phenomena into single elements. The scientific experiment is different from standard experiences in that it is guided by a mathematical theory that poses a question and is able to interpret the answer. It thereby transforms the given "nature" into a manageable "reality". Aristotle wanted to preserve Nature, to save the phenomena; his fault was in making too much use of common sense. Galileo dissects Nature, teaching us to produce new phenomena and to defy common sense with the help of mathematics (Von Weizsäcker 1996, pp. 104–105).

A more Platonic attitude can elevate the role of reason in the construction of idealized scientific models, beyond Aristotelian common-sense experience. Aristotle was able to express, in an abstract and consistent manner, many spontaneous perceptions concerning the universe that had existed for centuries before he gave them a logical and verbal rationale. In many cases, these were precisely the perceptions that since the 17th century, elementary education has increasingly banished from the Western mind. Today, the view of Nature held by most sophisticated adults shows few significant parallels to that held by Aristotle or even by the members of primitive collectives, yet views held by many non-Western peoples do parallel Aristotelian perspective with surprising frequency (Kuhn 1970, p. 95).

5 Conclusions

Post-positivist philosophers such as Kuhn, Feyerabend, and Dudley Shapere argued that the alleged distinction between theory and observation was to some extent illusory and untenable. The theory-ladenness of observation plays an important role in Galileo's justification of the heliocentric hypothesis by mechanics through thought experiments and thanks to astronomical observations through the telescope.

First of all the present study introduced abduction and idealization as processes devoted to the generation of new hypotheses. Koyrė (1978) regarded Galileo's new approach in building mathematical models of motion as the victory of a Platonic and abstract idealizing approach to science over that of Aristotle and medieval Aristotelians, who appealed to experience: "[F] or the contemporaries and pupils of Galileo, as well as Galileo himself, the Galilean philosophy of Nature, appeared as a return to Plato, a victory of Plato over Aristotle" (pp. 68–74). As Plato's student, Aristotle could have come close to understanding the principle of inertia, but he mainly focused on observation, turning away from Platonic idealizations such as those concerning eternity, vacuum, and the perfect, frictionless surfaces. He thereby lost the opportunity of discovering the concept of inertia (Potter 2009, pp. 82–83).

Today's teaching of science reflects yesterday's philosophy of science: several decades ago, empiricism dominated educational thought. It was taught that the scientist basically learns from experience by collecting data and generalizing or finding regularities in the data. The theoretical level is now more important even if we have to remember that the most important scientific theories did not fall from the sky: the views of discovery expounded in this article reflect the naturalistic turn in philosophy of science favored by the analysis of Galileo's thought experiments,strongly based on abstraction and idealization.

Acknowledgments This work was supported by the National Research Foundation of Korea Grant funded by the Korean Government (NRF2014S1A5B6037734).

References

Bhaskar, R. (1979/1998). *The possibility of naturalism: A philosophical critique of the contemporary social science*. London: Routledge.

Bhaskar, R. (1975/1997). *A realist theory of science*. London: Verso.

Brown, J. R. (1991). *The laboratory of the mind: Thought experiments in natural sciences*. New York: Rutledge.

Brown, J. R. (2000). Thought experiments. In W. H. Newton-Smith (Ed.), *A companion to philosophy of science*. Oxford: Blackwell Publishers.

Cartwright, N. D. (1989). *Nature's capacities and their measurement*. Oxford: Clarendon Press.

Chalmers, A. (1990). *Science and its fabrication*. Minneapolis: The University of Minnesota Press.

Cohen, I. B. (1985). *The birth of a new physics*. New York: W.W. Norton & Company Inc.

Cohen, I. B. (1950/1993). A sense of history in science. *Science & Education, 2*(3), 251–277 [1950, *American Journal of Physics, 18*, 343–359].

Cushing, J. T. (1998). *Philosophical concepts in physics: The historical relation between philosophy and scientific theories*. Cambridge: Cambridge University Press.

Dijksterhius, E. J. (1986). *The mechanization of the world picture, Pythagoras to Newton*. Princeton: Princeton University Press.

Drake, S. (1990). *Galileo: Pioneer scientist*. Toronto: University of Toronto Press.

Fernández-González, M. (2013). Idealization in chemistry: Pure substance and laboratory product. *Science & Education*, *22*(7), 1722–1740.

Feyerabend, P. K. (1975). *Against method: Outline of an anarchistic theory of knowledge*. London: New Left Books.

Finocchiaro, M. A. (2010). Defending Copernicus and Galileo: Critical reasoning and the ship experiment argument. *The Review of Metaphysics*, *64*, 75–103.

Galileo, G. (1967). *Dialogue concerning the two chief world systems*, Stillman Drake (trans.), Berkeley and Los Angeles: University of California Press.

Galili, I. (2009). Thought experiments: Determining their meaning. *Science & Education*, *18*, 1–23.

Garrison, J. W. (1986). Husserl, Galileo, and the process of idealization. Syntheses, 66, 329–338.

Gendler, T. S. (1998). Galileo and the indispensability of scientific thought experiment. *British Journal for the Philosophy of Science*, *49*, 397–424.

Gower, B. (1997). *Scientific method: A historical and philosophical introduction*. New York: Rutledge.

Koyrė, A. (1978). *Galileo studies*. Trans. J. Mepham, Humanities Press.

Kuhn, T. (1970). *The structure of scientific revolution* (2nd ed.). University of Chicago Press: Chicago.

Ladyman, J. (2002). *Understanding philosophy of science*. London: Routledge.

Losee, J. (2001). *A historical Introduction to the philosophy of science* (4th ed.). New York: Oxford University Press Inc.

Magnani, L. (2001). *Abduction, reason, and science. Processes of discovery and explanation*. New York: Kluwer Academic/Plenum Publishers.

Magnani, L. (2009). *Abductive cognition: The epistemological and eco-cognitive dimensions of hypothetical reasoning*. Berlin: Springer.

McMullin, E. (1985). Galilean idealization. *Studies in History and Philosophy of Science*, *16*(3), 247–273.

Nola, R. (2004). Pendula, models, constructivism and reality. *Science & Education*, *13*, 349–377.

Nowak, L. (1994). Remarks on the nature of Galileo's methodological revolution. In M. Kuokkanen (Ed.), *Idealization VII: Structuralism, idealization and approximation: Pozna'n studies in the philosophy of science and the humanities* (Vol. 42, pp. 111–126). Amsterdam: Rodopi.

Norton, J. D. (1996). Are thought experiments just what you thought? *Canadian Journal of Philosophy*, *26*, 333–366.

Oh, J.-Y. (2012). Understanding scientific inference in the natural sciences based on abductive inference strategies. In L. Magnani & P. Li (Eds.), *Philosophy and cognitive science: Western & Eastern studies*, SAPERE 2 (pp. 221–237). New York: Springer.

Oh, J.-Y. (2014). Understanding natural science based on abductive inference: Continental drift. *Foundations of Science*, *19*(2), 153–174.

Parisi, A. (2001). How did universal attraction force be discovered? Translated by J. H. Ahn as Korean Language. LAPIS, Rome.

Portides, D. P. (2005). A theory of scientific model construction: The conceptual process of abstraction and concretization. *Foundations of Science*, *10*, 67–88.

Potter, C. (2009). *You are here: A portable history of the universe*. New York: Harper Collins Publishers.

Psillos, S. (2007). *Philosophy of science A-Z*. Edinburgh: Edinburgh University Press.

Rescher, N. (1978). *Peirce's philosophy of science*. Indiana: University of Notre Dame Press.

Suppe, F. (1989). *The semantic conception of theories and scientific realism*. Urbana: University of Illinois Press.

Von Weizsäcker, C. F. (1996). *The relevance of science*. UK: HarperCollins Publisher.

Whewell, W. (1847). *The philosophy of the inductive sciences*. London: John W. Parker.

Biomorphism and Models in Design

Cameron Shelley

Abstract Biomorphism is a form of biomimicry that involves the use of biological forms as models for the design of artifacts such as airplanes, computers, and islands. This article characterizes biomorphism as a form of abductive reasoning. It also provides an overview of biomorphic design in terms of the parameters of similarity and utility. The cognitive significance of biomorphism is reviewed with respect to research in pareidolia and consumer choice. The normative status of biomorphism is considered in light of its tendency to conflate natural and artificial categories.

1 Ahead with Macintosh

The form of the original Macintosh computer was quite different from other computers at the time it was designed, in 1981. At that time, a home computer was much like a TV set perched on a box that contained the CPU. Steve Jobs wanted the Macintosh to be different. Also, he wanted it to look friendly instead of aloof or threatening. He accomplished this in a way that was both subtle and somewhat surprising (Isaacson 2011, p. 129):

> Jobs kept insisting that the machine should look friendly. As a result, it evolved to resemble a human face. With the disk drive built in below the screen, the unit was taller and narrower than most computers, suggesting a head. The recess near the base evoked a gentle chin, and Jobs narrowed the strip of plastic at the top so that it avoided the Neanderthal forehead that made the Lisa subtly unattractive.

The idea that a person would find it pleasant to have a disembodied head in their desk seems implausible. However, there is a kind of sense to it. If a computer is viewed as a kind of thinking machine, then it might be appropriate for it to look like it has a face, and that people should read the face as a part of the process of interacting with it (Kunkel 1997, p. 26):

C. Shelley (✉)
Centre for Society, Technology and Values, University of Waterloo,
Waterloo, ON, Canada
e-mail: cam_shelley@yahoo.ca

L. Magnani et al. (eds.), *Philosophy and Cognitive Science II*,
Studies in Applied Philosophy, Epistemology and Rational Ethics 20,
DOI 10.1007/978-3-319-18479-1_12

Fig. 1 A Macintosh 128 K, courtesy of the *All about Apple* museum, via Wikimedia commons

> The idea of a computer as a head on a desktop, with a face and a chin, encourages the user to think of it as an alter ego, a desktop friend that will always be there.

In other words, the face-like appearance of the Macintosh might help users to relate to the machine in a way that is appropriate for a device whose job is that of a cognitive assistant (cf. Shelley 2015). See Fig. 1.

2 Biomorphism

Designing a computer to resemble a head is an example of *biomorphism*. In general terms, biomorphism is a kind of biomimicry that refers to the imitation of biological structures or systems in artificial devices. Biomorphism can be quite literal, as it often is in art or sculpture, or more abstract, as in the case of the original Macintosh. It may be used for various purposes, e.g., aesthetic or functional. In the case of the Macintosh, both these purposes appear to be intended. The face-like appearance of the machine was intended by Jobs to be perceived as pleasing and friendly. In addition, it was intended to suggest to users that the machine would function as a kind of companion or assistant, and not merely as a passive tool like a TV set or a hammer.

Although biomorphism in design has been practiced and promoted in various ways, its logic has not been explored in a systematic way. The purpose of this article is to describe some parameters of biomorphism in design in terms of abductive reasoning. In a technical sense, abductive reasoning is the reasoning of explanation, e.g., the setting out and testing of hypotheses. However, this linguistically-based view has been extended to other cognitive modalities, such as models and visual representations (cf. Magnani 2001). Because of its emphasis on appearances, an account of biomorphism in design will call upon the broader concept of abductive reasoning.

In this article, I will sketch out a description of biomorphism in design by looking at its different manifestations. Then, I will outline an account of biomorphism in design as a sort of abductive reasoning. In so doing, I aim to clarify what biomorphism in design is, and how it may be done well or poorly. This result should also provide an example of how abductive reasoning is not only a matter of hypothesis generation and testing in a scientific mode.

3 Design Space of Biomorphism

There is not simply one sort of biomorphism. Instead, there are a variety of ways in which biological form, animal, vegetable, or otherwise, gets embodied in designs. These ways depend upon the role that biomorphism plays in design. So, to come to grips with biomorphism, we need some paradigm for understanding the kinds of roles that it can play. Here, I propose a basic parameterization of the roles of biomorphism to provide a start on this project.

I propose to use two parameters for this purpose. The first is a dimension of *similarity*. That is, biomorphism may be more or less literal in appearance. An *abstract* use of biomorphism is one in which a biological form is not used in a concrete or realistic way. The Macintosh would be a good example of an abstract use of biomorphism. The head-like appearance of the Macintosh is not evident to a superficial examination of the design. Instead, it becomes apparent only when elements that are borrowed from a head are pointed out, after which the total biomorphism can be inferred.

The opposite of biomorphism based on an abstract borrowing from animal form is a literal borrowing, or a *resemblance*. Classical sculpture provides many good examples of biomorphism of resemblance. The Laocoön group, for example, purports to show the Trojan priest Laocoön who is being strangled by serpents, an episode from Greek literature (cf. Boardman 1993). See Fig. 2. The sculpture performs this function by providing realistic-looking figures representing Laocoön, his two sons, and the serpents sent by Poseidon to strangle them. The identity of humans and animals is suggested by carvings that resemble what viewers would see if those beings were actually before their eyes.

Fig. 2 The Laocoön group, courtesy of Sailko via Wikimedia commons

Fig. 3 The Palm Jumeriah, courtesy of NASA via Wikimedia commons

Besides abstraction, another parameter of biomorphism is *utility*. Utility identifies the kind of work that biomorphism is supposed to perform. Following Heskett (2002), we can think of designs as doing at least two sorts of work. The first may be called *function*. A purely functional design is one that is designed to allow users to effect some physical change in the world. No other consideration is relevant. An airplane provides a good example. The aerodynamic shape of the body and wings of an airplane recall those of a bird. In early airplane design, resemblance to birds was pursued for its functional benefits, so that early aviators could understand the problems of lift and propulsion that bird form addressed in nature (cf. Spenser 2008).

The opposite of biomorphism based on function of animal forms is biomorphism based on the *significance* of animal forms. Here, significance refers to the social or cultural associations that animal forms have for people. As Heskett points out, these associations give meaning to designs beyond their functional possibilities. For example, the so-called Palm Islands of Dubai are a set of islands in the Persian Gulf formed to resemble palm trees when seen from the air. See Fig. 3. In functional terms, this structure helps to provide a great deal of beachfront property for owners of housing, hotels, resorts, and theme parks on the islands. In additional to this function, the resemblance of the archipelago to palm trees has cultural significance. The palm tree is a national symbol of Dubai and palm fronds were used in the construction of traditional dwellings in the region (Molavi 2007). In addition, the palm tree has international significance as a facet of exotic tropical islands, or oases in the desert. Each of these associations provides cultural and commercial significance to the design of the islands.

Together, the dimensions of similarity and utility divide up the space of biomorphisms in design in a way that aids us in understanding how they might work.

As with any paradigm applied to a complex phenomenon, there are problem cases. There are designs, for example, that seem to occupy more than one point in this space. The Flying Tigers, a group of American volunteers who fought the Japanese in China early in World War Two, were famous for the shark face that they painted on the front of their P 40 fighter planes (Clements 2001). See Fig. 4. As such, the planes were biomorphic in two ways: They bore a functional similarity to birds on the one hand,

Fig. 4 A P-40 fighter plane, courtesy of the National Museum of the U.S. Air Force photo 050215-F-1234P-065 via Wikimedia commons

and also a symbolic similarity to sharks on the other hand. Thus, their biomorphism may seem to be confound categorization.

This problem is resolved when we observe that the planes of the Flying Tigers play these different roles in different ways. They are functionally like birds due to their overall aerodynamic shape, whereas they are significantly like sharks due to their decoration. So, it is through these different aspects of design that the planes play two roles. We can avoid confusion by bearing in mind that the assignment of designs to roles is in virtue of such aspects of design. As a result, one object can be assigned to multiple roles without contradiction.

The biomorphic design space described by these dimensions can be visualized as in Fig. 5. The utility of the head-shape of the Macintosh has been classed as *significant* because its use was realized through the associations that users would have with heads. Similarly, the utility of the Laocoön sculpture has been classed as *significant* because its use was primarily to evoke associations with the myth.

The similarity of the airplane has been classed as *abstract* because it has only a rough likeness to an actual bird. The similarity of the Palm islands to actual palm trees has been classed at some point intermediate between the realism of the Laocoön sculpture and the abstractness of the Macintosh. The islands have an approximate resemblance to a palm tree, taken in profile, which is nevertheless not as subtle as the resemblance of the computer to an actual head.

The additional example of a *glove* illustrates a design that resembles part of an organism and whose utility is to do work similar to that part. A glove is something that resembles a hand, by definition. For some gloves, such as fingerless ones, the

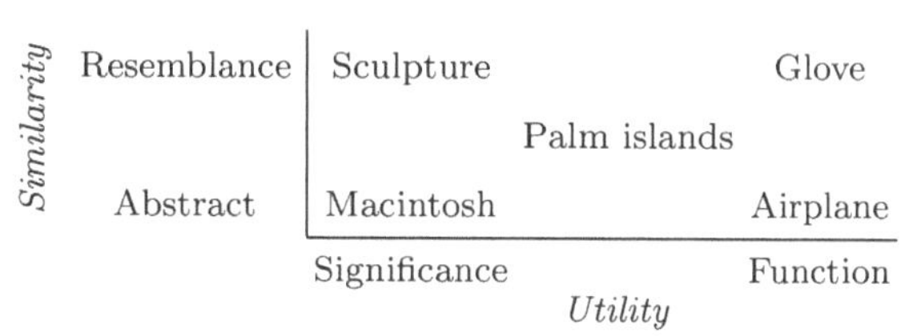

Fig. 5 Biomorphic design broken into similarity and utility dimensions

resemblance is weaker than for standard gloves. The shape and covering of the glove have functions similar to the shape and function of the skin of the human hand, namely to facilitate touching and grasping, and to keep the inner workings safe from cold or injury. In other words, the function of a glove is to act like a second skin for a human hand, a function that it realizes partly through biomorphic design.

4 Visual Abduction and Biomorphism

At this point, we can relate biomorphism to visual abduction. Abduction is a form of reasoning usually associated with explanation. That is, a typical example of abduction is an inference that a draft of air in a room is due to the presence of an open window in the house. Because they are explanatory, abductive inferences are not certain. For example, a draft of air in a room could be due to a fan and not to an open window. There is a large body of research in abductive reasoning concerning what abductions are, what distinguishes good abductions from others, and how abductive reasoning fits into the broader ecology of cognition (cf. Magnani 2009).

Visual abduction concerns abductive reasoning in which visual mental representations are active or predominant. In other words, visual abduction is abductive reasoning in which the minds eye plays a significant role. For example, seeing a cat stuck up in a tree (e.g., Fig. 6) might prompt someone to imagine the route the cat took to get there and what caused it to do so: pursuing a squirrel or being chased by a dog, for example. Even visual perception has an abductive character. For example, when you see a car parked behind a pole (e.g., Fig. 7), you see only two objects, namely the car and the pole, and not three objects, namely, the pole and two half-cars with a space between them hidden behind the pole. The perception of a single car is a kind of abduction: The presence of only one car explains why it appears that the front and rear ends of a car emerge from behind either side of the pole.

Model-based reasoning can be an important form of abduction. In such reasoning, information about one thing is supplied from something else. A person reasoning

Fig. 6 A cat up in a tree, courtesy of Broc via Wikimedia commons

Fig. 7 A truck parked behind a pole; photo by author

about how a cat got stuck in a tree might draw upon a time when a different cat got stuck in the same tree, or in a similar tree elsewhere, for example.

Biomorphism in design is another illustration of model-based reasoning. That is, the designer of a biomorphic object uses the form of an organism as a model for the form of an artifact. As noted earlier, Steve Jobs used the human head as a model for the form of the original Macintosh case. In addition, biomorphism is frequently visual insofar as it is based upon aspects of the visual appearance or layout of an organism being used as the model for an artifact.

Model-based design is not an inference. That is, an organism that inspires a biomorphic artifact is not an explanation of that artifact. However, the modeling in design does have a hypothetical quality that it shares with abductive inference. When using a model, a designer selects attributes of the model to embody in the artifact. The selection is done on the basis that the selected attributes will help the designer to create a design that solves a given problem. The designer cannot be certain, however, that the selected attributes are either relevant to the problem at hand or sufficient to solve it.

In the case of the original Macintosh, Jobs selected only some attributes of a human head to transfer to the case, namely its proportions and the arrangement of features in a face projecting forward from the case. He could not be certain that those features would indeed help to make the computer seem friendly to users, nor could he be certain that some features that he omitted would not also be helpful.

Biomorphism in design can therefore be treated as a form of model-based, abductive reasoning that has a particular, visual characteristic.

5 Biomorphism and Cognition: Pareidolia

In order to know how designers can use biomorphism to solve design problems, we need to know how biomorphism is perceived by users of their designs. There has been no general work in this area. However, there has been a little research into some specific cases that are relevant to the present topic.

Fig. 8 The so-called Mars Face, courtesy of Nasa photo PIA01141 via Wikimedia commons

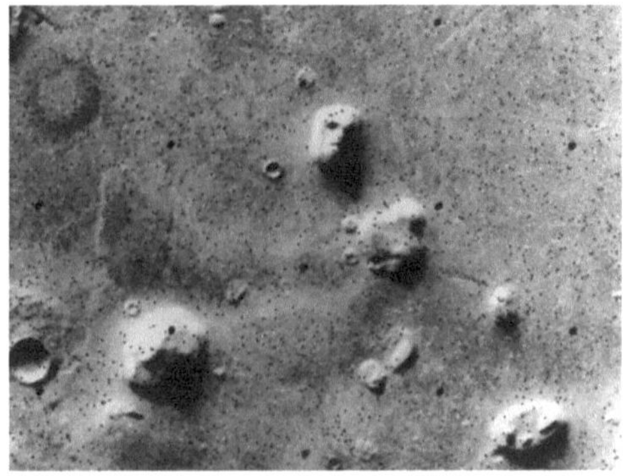

The first body of relevant research is into a psychological phenomenon called *pareidolia*. Pareidolia refers to the tendency of people to see significant patterns in vague or random stimuli (cf. Chalup et al. 2010). The classic example of pareidolia is the perception of faces in non-facial representations. Very often, the face perceived is one of cultural significance, e.g., the face of Jesus. For example, patrons of a Tim Horton's donut shop in Bras D'Or, Nova Scotia noticed a stain on a wall that resembled the face of Jesus. The incident received broad press coverage and was later the basis for a play called "Halo" and a movie called *Faith, fraud, and minimum wage* (Grant 2013).

Another example would be the famous Mars face which was noticed by a Mars photo enthusiast Vincent DiPietro in 1977 among Viking images of the Martian surface taken the year before. See Fig. 8. The photo seems to show a human face detailed in relief about a mile wide. DiPietro and others speculated that the face was a deliberate design left by authors unknown (Gardner 1985). Later and more detailed photos of the region did not show the same resemblance.

It is plausible to think that pareidolia might be a basic trait of human cognition. Carl Sagan speculated that it would be adaptively advantageous for early humans if they were more likely to mistake non-facial images for faces than the reverse (Sagan 1995). After all, the cost of overlooking the face of an enemy or predator would be more than the cost of thinking one is there when it is not.

Neuroimaging research has suggested that there is a specific area of the brain, known as the Face Fusiform Area (FFA), that is specialized in the recognition of faces (Kanwisher et al. 1997). The FFA appears to be particularly active during face recognition, and at an early stage in the processing of facial representations.

Interestingly, a recent study also suggests that the FFA is similarly active in the processing of pictures of artifacts that are face-like in appearance (Hadjikhani et al. 2009). It was found that these objects provoked responses in test subjects similar to responses to pictures of human faces with neutral expressions. The authors drew two main conclusions from their study. First, the processing of these stimuli as faces seems to occur quite soon after presentation, at about 165 ms. Thus, it seems that peoples brains are wired to find faces in visual data by default. Second, the resemblance to faces could be quite crude, often requiring only representations of two eyes and a

Fig. 9 Former Brazilian socket type, courtesy of Sylx100 via Wikimedia commons

nose or mouth. Think of emoticons such as the so-called smiley, :-) or the socket in Fig. 9. The outline of a face did not have to be present for subjects to agree that the image seemed face-like, and to provoke the FFA.

Research into pareidolia, then, supports the suggestion that biomorphism may have an immediate and powerful impact on users of artifacts so designed. The nature of that impact remains somewhat obscure, however.

6 Biomorphism and Cognition: Consumer Choice

There has been some research into how anthropomorphism in design affects consumer preferences for products. Anthropomorphism is a particular form of biomorphism in which the human form is used as a model for design. Recent work suggests that people have a robust preference for designs that appear human-like in some way, e.g., cars that have a face-like front profile.

This preference depends on several contextual factors (Miesler 2012). Miesler compared the preferences of test subjects between two groups of cars, with more face-like or more neutral front profiles against a variety of consumer goals. See Fig. 10 for the front profiles.

In the end, Miesler found that subjects do tend to prefer anthropomorphic profiles when they are asked to consider emotional goals of car ownership such as trust in the car, pride of ownership, or commitment to long-term ownership. In other words, anthropomorphic appearance seemed more positive when the person's relation to the car is conceived in social or emotional terms.

However, when the relationship is conceived in more instrumental terms, the preference was reversed. That is, when people were asked for their preference in cars based on considerations such as efficiency, robustness, or economy, they tended to prefer cars with neutral front profiles. The sizes of both effects were also correlated with the subjects' sensitivity to the anthropomorphic quality of the cars' profiles.

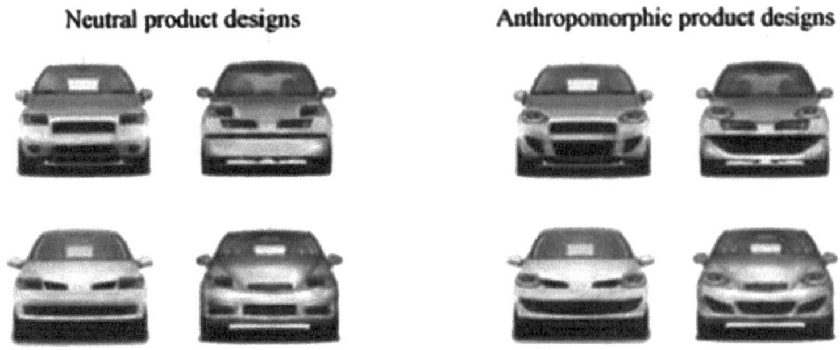

Fig. 10 Neutral and anthropomorphic car fronts, courtesy of (Miesler 2012)

It appears, then, that biomorphism can reinforce a tendency of people to think of their relationships with artifacts in interpersonal terms, provided they are already inclined to do so. Under those circumstances, anthropomorphic appearance, if it is pleasant, can prompt users to relate to their belongings in terms of trust, pride, and commitment.

Although Miesler's research does not bear on this point, it is a rule of thumb in car design that the front profile of a car can put off drivers if its face appears misshapen. Car critic Neil (2007), for example, holds that the 1998 Fiat Multipla suffers from this problem. The appearance of extra, high-beam bulbs at the base of the windshield made it look like "it had several sets of eyes, like an irradiated tadpole. I rented one of these in Europe and it worked beautifully, but it was just so tragic to look at."

Another possibility not tested in this research is that biomorphism could be used to affect people's judgments about the functional qualities of a design. For example, car bodies are sometimes designed to be "muscular", reminiscent of powerful animals such as cheetahs or jaguars. The assumption by the designers is that people will transfer their impression of power from the animal to the vehicle. The idea seems plausible but needs to be tested.

There is some empirical support for the claim that biomorphism can affect how people judge designs. Existing research is not complete and also suggests that the impact of biomorphism is complicated by the context in which designs are assessed.

7 Biomorphism and Honesty

Accounts of visual abduction (or abductive reasoning of any sort) seek not only to describe its cognitive aspects but also to understand what distinguishes good reasoning from poor reasoning. That is, what makes an abduction a good one or not? One approach is to assess the rationality of reasoning, that is, to assess how well abductions are logically supported by evidence.

In design, a normative approach to the assessment of abductive reasoning is also appropriate. That is, it is appropriate to assess designs against norms of socially acceptable conditions, such as privacy laws and safety regulations. For biomorphism, normative issues concern how well biological models in design comport with such norms. One norm of good design that raises an issue for biomorphism is *honesty*.

Concerns about honesty in design began in the 19th century and continue to be expressed in the present day. For example, the eminent industrial designer Dieter Rams made honesty the sixth of his Ten Commandments of Good Design (Lovell 2011, p. 354). Although expressions of this principle vary, it can be summarized briefly as follows. An honest design (Shelley 2015):

1. Does not disguise what it is, and
2. Exhibits what it is.

On this principle, for example, a building made of concrete and steel should exhibit these materials, instead of covering them up with bricks so as to look more traditional that it truly is.

Honesty appears to be an obstacle where biomorphism is concerned. The problem arises because biomorphism, by definition, means applying organic forms to non-organic artifacts. For a simple example, consider cell phone towers in southern California that are made to look like palm trees. See Fig. 11. The purpose of disguising the towers is clear enough: It is intended to help them look innocuous in urban areas where industrial equipment would otherwise be an unwelcome intrusion (Barratt and Whilelaw 2011). It may seem that disguise is necessary in this case. However, a designer like Dieter Rams might argue that disguise is just an excuse for poor design: A thoughtfully designed tower would look good without the need for subterfuge.

Fig. 11 Cell phone tower disguised as a palm tree, courtesy of Gary Minnaert via Wikimedia commons

The view that good design requires honesty presents a challenge for biomorphism in design. It seems inherently to involve a degree of pretense that an artifact can have a natural, biological form. There will be occasions where there is no difficulty, e.g., with the aerodynamic form of airplanes and birds. In such cases, the nature of the problem constrains the forms of both organisms and artifacts. However, it is not clear that such constraints apply to items such as desktop computers like the Macintosh. Computing does not seem inherently to require a head-like form.

8 Conclusions

Biomorphism is an interesting form of model-based reasoning applied to design. It involves the use of biological forms as models for the design of artifacts. As an issue of outer form, biomorphism is also inherently visual in nature. Thus, biomorphism in design can be considered as a kind of visual abductive reasoning.

To understand biomorphism as a form of visual abduction, we need to identify its parameters. As a beginning, two relevant parameters discussed here are:

1. Similarity: the literalness or abstractness of the resemblance between model and design, and
2. Utility: the extent to which the model is employed for social or physical effect.

Together, these parameters define a space of biomorphic design that helps to understand what is going on in any particular case. Designs can be located in any point in this space, or even several points where different models are employed in one result.

The cognitive impact of biomorphism is supported by at least two threads of psychological research. The phenomenon of pareidolia, the perception of significant form in vague or random data, confirms that people can be highly attuned to the presence of even highly abstract natural form in artifacts. Research into consumer choice suggests that such perceptions affect how people judge designs. Biomorphic design can supply people with cues that then affect how people assess the functionality and likeability of artifacts such as cars. Failure to take these factors into account can adversely affect how designs are assessed.

Of course, biomorphism must be treated with caution. Since biomorphic design typically involves the conflation of natural and artificial forms, it can be considered dishonest and therefore poor work. Disguising a cell phone tower as a palm tree could, for instance, be regarded as a cover up for a half-hearted design effort, rather than an appropriate solution adapted from nature. On the whole, biomorphism is an interesting design paradigm and one that deserves to be further studied and more widely understood.

References

Barratt, C., & Whilelaw, I. (2011). *A spotter's guide to urban engineering: Infrastructure and technology in the modern landscape*. Buffalo, NY: Firefly Books.

Boardman, J. (1993). *Oxford history of classical art*. Oxford: Oxford University Press.

Chalup, S. K., Hong, K., & Ostwald, M. J. (2010). Simulating pareidolia of faces for architectural image analysis. *International Journal of Computer Information Systems and Industrial Management Applications*, *2*, 262–278.

Clements, T. J. (2001). *American Volunteer Group colours and markings*. Oxford: Osprey.

Gardner, M. (1985). The great stone face and other nonmysteries. *Skeptical Inquirer*, *10*(2), 14–18.

Grant, L. J. (2013). Film was inspired by a vision of Jesus at Tim Hortons. Retrieved October 18, 2013 from http://www.capebretonpost.com/Living/2010-11-30/article-2006957/Film-was-inspired-by-vision-of-Jesus-at-Tim-Hortons/1, Nov 2010.

Hadjikhani, N., Kestutis, K., & Naik, P. A. (2009). Early (m170) activiation of face-specific cortex by face-like objects. *Neuroreport*, *20*(4), 403–407.

Heskett, J. (2002). *Toothpicks and logos: Design in everyday life*. Oxford: Oxford University Press.

Isaacson, W. (2011). *Steve Jobs*. New York, NY: Simon & Schuster.

Kanwisher, N., McDermott, J., & Chun, M. N. (1997). The face fusiform area: A module in human extrastriate cortex specialized for face perception. *Journal of Neuroscience*, *17*(11), 4302–4311.

Kunkel, P. (1997). *Appledesign: The work of the Apple industrial group*. New York, NY: Watson-Guptill.

Lovell, S. (2011). *Dieter Rams: As little design as possible*. London: Phaidon.

Magnani, L. (2001). *Abduction, reason, and science: Processes of discovery and explanation*. New York, NY: Kluwer Academic Publishers.

Magnani, L. (2009). *Abductive cognition: The epistemological and eco-cognitive dimensions of hypothetical reasoning*. Berlin: Springer.

Miesler, L. (2012). Product choice and anthropomorphic designs: Do consumption goals shape innate preferences for human-like forms? *The Design Journal*, *15*(3), 373–392.

Molavi, A. (2007, January). *National Geographic* Dubai: Sudden city.

Neil, D. (2013). The 50 worst cars of all time: 1998 Fiat Multipla. Retrieved October 18, 2013 from http://content.time.com/time/specials/2007/article/0,28804,1658545_1658544_1658537,00.html, Oct 2007.

Sagan, C. (1995). *The demon-haunted world-science as a candle in the dark*. New York, NY: Random House.

Shelley, C. (2015). The nature of simplicity in Apple design. *The Design Journal*, *18*(3).

Spenser, J. (2008). *The airplane: How ideas gave us wings*. New York, NY: Harper Collins.

MIX
Papier aus verantwortungsvollen Quellen
Paper from responsible sources
FSC® C105338
FSC
www.fsc.org

If you have any concerns about our products,
you can contact us on
ProductSafety@springernature.com

In case Publisher is established outside the EU,
the EU authorized representative is:
Springer Nature Customer Service Center GmbH
Europaplatz 3, 69115 Heidelberg, Germany

Printed by Libri Plureos GmbH
in Hamburg, Germany